AIMING FOR AN A IN A-LEVEL GEOGRAPHY

Simon Oakes

HODDER EDUCATION
AN HACHETTE UK COMPANY

Acknowledgements

With thanks to the CDARE team at the Sheffield Institute of Education for their assistance in developing and reviewing this title.

Every effort has been made to trace all copyright holders, but if any have been inadvertently overlooked, the Publishers will be pleased to make the necessary arrangements at the first opportunity.

Although every effort has been made to ensure that website addresses are correct at time of going to press, Hodder Education cannot be held responsible for the content of any website mentioned in this book. It is sometimes possible to find a relocated web page by typing in the address of the home page for a website in the URL window of your browser.

Hachette UK's policy is to use papers that are natural, renewable and recyclable products and made from wood grown in sustainable forests. The logging and manufacturing processes are expected to conform to the environmental regulations of the country of origin.

Orders: please contact Bookpoint Ltd, 130 Park Drive, Milton Park, Abingdon, Oxon OX14 4SE. Telephone: (44) 01235 827827. Fax: (44) 01235 400401. Email: education@bookpoint.co.uk. Lines are open from 9 a.m. to 5 p.m., Monday to Saturday, with a 24-hour message answering service. You can also order through our website: www.hoddereducation.co.uk

ISBN: 978 1 5104 2408 1

First published in 2018 by

Hodder Education
An Hachette UK Company
Carmelite House
50 Victoria Embankment
London EC4Y 0DZ

www.hoddereducation.co.uk

Impression number 10 9 8 7 6 5 4 3 2 1
Year 2021 2020 2019 2018

Typeset by Integra Software Services Pvt. Ltd., Pondicherry, India

Printed in Spain

A catalogue record for this title is available from the British Library.

Contents

Getting the most from this book

Aiming for an A is designed to help you master the skills you need to achieve the highest grades.

The following features will help you get the most from this book.

Learning objectives

> **A summary of the skills that will be covered in the chapter.**

Exam tip

Practical advice about how to apply your skills to the exam.

Activity

An opportunity to test your skills with practical activities.

Common pitfall

Problem areas where candidates often miss out on marks.

The difference between...

Key concepts differentiated and explained.

Annotated example

Exemplar answers with commentary showing how to achieve top grades.

Worked example

Step-by-step examples to help you master the skills needed for top grades.

Take it further

Suggestions for further reading or activities that will stretch your thinking.

You should know

> **A summary of key points to take away from the chapter.**

About this book

The skills roadmap

Geography A-level is a popular subject whose diverse and fascinating content ranges from the study of changing places to the operations of the carbon cycle. The sheer breadth of included topics (with accompanying skills and fieldwork experiences) makes geography an attractive choice for many learners — especially those students who enjoyed the breadth of the GCSE curriculum and view themselves as being academic 'all-rounders'. But geography's expansive knowledge and skills demands can feel challenging at times too. A-level geographers are required to:

- engage with a range of scientific, political, social and philosophical ideas
- become competent statisticians and data analysts
- write persuasive and argumentative essays

This book serves as a supporting 'roadmap' for the study of A-level Geography by guiding you towards the different study and assessment goals that lie ahead.

- Chapter 1 focuses on how to read and engage with diverse literatures in an active and critical way during your 2-year course. There is more to reading than some people think.
- Chapter 2 helps you develop a broad analytical skill set and become proficient at completing and commenting on statistical tests.
- Chapter 3 shows you how to tackle short-answer tasks, either as part of your ongoing studies or in the final examinations. There is more of an art to concise and directed writing than many students at first realise.
- Chapter 4 deals with evaluative essay writing, which is the greatest challenge posed by the course for many students (particularly those who are not studying other subjects at A-level that require them to develop evaluative essay-writing skills). This chapter explores ways of structuring an argument and arriving at an evidenced conclusion that includes a substantiated judgement. If you are not sure what is meant by 'evidenced' or 'substantiated', then this is definitely a chapter that you will want to read closely.
- Chapter 5 focuses on the independent investigation (sometimes called the non-examined assessment, or NEA). There are important steps that you can actively take towards maximising your NEA marks and boosting your overall chances of gaining an A/A* grade.
- Chapter 6 provides some final exam-board-specific revision advice and looks in turn at the unique characteristics of the AQA, Edexcel, Eduqas, OCR and WJEC courses and assessments, with an eye to achieving the highest grades.

You first need to learn about the assessment objectives (AOs) that provide the framework for your course and its accompanying examined and non-examined assessment components (the latter being a 4000-word independent investigation, which is worth 20% of your total mark).

Assessment objectives, synopticity and levels-based marking

A-level study allows you to focus on a small handful of subjects that you really enjoy; and enjoyment is the key to success. In general, people become more highly motivated to work when they like what they are doing. During your 2 years of A-level study, seize every opportunity you can to read widely and watch informative documentaries that interest you. Keep an eye on current affairs, such as global political events and climate change updates. However, there is no escaping the fact that everything you do leads ultimately to an examination that you want to succeed in. Knowing as much as possible about the assessments — including how they are designed and what skills they are testing (other than subject knowledge) — is essential.

Table 1 Assessment objectives for A-level Geography

	Description	Relation to exam command words
AO1	Demonstrate knowledge and understanding of places, environments, concepts, processes, interactions and change, at a variety of scales	These AO1 exam command words invite you to share your knowledge about something with an examiner. **State** Give a specific brief answer or name without explanation. **Describe** Give a detailed account. **Outline** Give a brief account or summary. **Explain** Give a reasoned, causal explanation of something (a standalone task involving recalled knowledge and understanding only).
AO2	Apply knowledge and understanding in different contexts to interpret, analyse and evaluate geographical information and issues	These command words additionally require you to *use* your knowledge in unexpected ways. You may need to offer the examiner an explanation of something new, or to make a judgement about an issue. **Explain** Give a reasoned, causal explanation of why something occurs (in the context of an unfamiliar scenario, such as a previously unseen question paper map, chart or other resource). **Suggest** For an unfamiliar scenario, provide a reasoned explanation of how or why something has occurred. **Assess** Use evidence to determine the relative significance or importance of something, having considered a range of relevant information. **Evaluate** Measure the value, importance, success or role of something and ultimately conclude by making a balanced and substantiated judgement. **Discuss** Offer a considered and balanced review that includes a range of arguments, factors or hypotheses, supported by appropriate evidence. **To what extent?** Consider the merits or otherwise of an argument or idea. Exam boards vary in their approach to using command words, so you should read Chapter 6 carefully in order to find out more about this.
AO3	Use a variety of relevant quantitative, qualitative and fieldwork skills to: investigate geographical issues; interpret, analyse and evaluate data and evidence; construct arguments and draw conclusions	Approaches vary among exam boards (see Chapter 6), but as a general rule: • Some AO3 command words require you to demonstrate competence in a mathematical or analytical skill (including **calculate** and **plot**). • Some AO3 command words require you to draw together lots of new information in order to construct arguments. The most widely used of these is **analyse**, which means 'use geographical skills to investigate information/issues in a structured or systematic way'. Your independent investigation is regarded as a mostly AO3 task.

The assessment objectives (AOs) for geography are the targets (objectives) against which you will be judged (assessment). All exam boards use the same geography AOs, which are shown in the table on page 6. Different questions in your examination will be linked with particular assessment objectives (or a combination of them). Commonly used question command words typically linked with each AO are shown in the table too. The exact approach taken differs among the exam boards, however; this is addressed at various points throughout this book and in Chapter 6 in particular.

The AOs for all your GCSE and A-level subjects have been designed and refined over time by education experts. Bloom's taxonomy (see below) was created in order to promote higher forms of thinking in education, such as analysing and evaluating concepts, processes, procedures and principles (rather than just remembering facts). 'Remember' and 'understand' are typically characterised as *lower-order* or *core* skills, whereas 'apply', 'analyse', 'evaluate' and 'create' are described as *higher-order* skills, requiring you to use knowledge in ways that 'add value' when carrying out an academic task.

Some people believe that the value attached to our ability to apply knowledge has increased since the advent of the internet. Now that people can look up facts easily online, there is arguably less need than there used to be to memorise encyclopaedic blocks of knowledge. Instead, society increasingly values someone's ability to critically analyse and evaluate the streams of information flowing past their fingertips. It is important that you can spot 'fake news' when carrying out research, for instance (page 16).

Note that 'create' appears in Bloom's taxonomy but not in the A-level Geography AOs and command words. However, a successful A-level independent investigation will undoubtedly be creative — insofar as you are expected to devise a unique research question with a bespoke methodology; and you are also expected to analyse, evaluate and reflect on the unique data set that you have created.

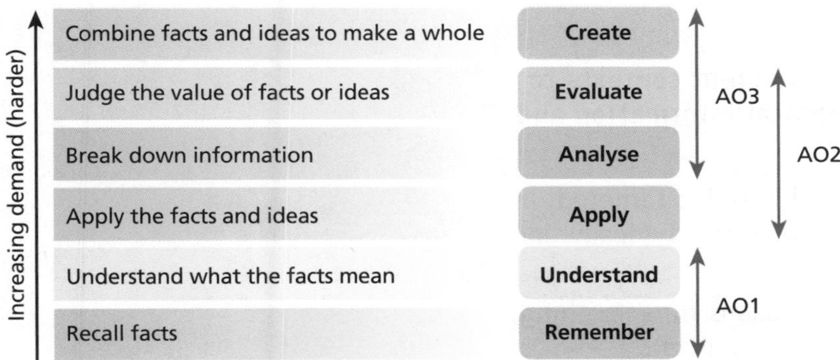

Figure 1 Bloom's taxonomy and the assessment objectives for geography

The majority of revision guides and textbooks on the market aim mainly to fulfil student and teacher demand for material that helps learners understand and remember ideas and case studies (higher-order 'assess' and 'evaluate' skills may be catered for partially by case studies that explore the costs and benefits of management, or the strengths and weaknesses of theories and models). This book complements these at-times-descriptive texts by providing hands-on examples of how geographical ideas, information and examples can be used to properly assess and evaluate in the context of the AOs.

Assessment objective 1 (AO1)

Demonstrate knowledge and understanding of places, environments, concepts, processes, interactions and change, at a variety of scales.

- This AO is about the facts, ideas and places you have learned about (this could be a case study of a water management scheme, or an example of a transnational company's world operations).
- It is important to be aware that there is a difference between *knowing* something and *understanding* it. AO1 rewards both 'knowledge' and 'understanding'. Some students will show a lack of understanding when writing out the facts about a physical process, for example.

The difference between...

Knowing	Understanding
Knowing something means that you are able to remember and describe key facts, such as migration push and pull factors, different weathering processes or the names of important hydrological processes. For example, the following shows knowledge: 'Freeze–thaw is a weathering process that operates in cold environments. Each night, freezing water expands in cracks. After many freezing cycles, this breaks apart chunks of rock.'	Understanding means you know more than just the basics. You also have a deeper and more secure understanding of *how* or *why* geographical processes operate. For example, a possible *lack* of understanding is demonstrated by the following: 'The colder the temperature becomes, the more effective freeze–thaw weathering will be.' Can you see why this shows knowledge *but a lack of understanding*?

AO1 tasks usually take the form of standalone, short-answer questions. Relatively few of these feature in A-level exams compared with GCSE exams. Chapter 3 looks at examination technique for answering these questions.

Assessment objective 2 (AO2)

Apply knowledge and understanding in different contexts to interpret, analyse and evaluate geographical information and issues.

This AO covers a wide range of different skills and is targeted by a variety of different tasks and command words. Typically, in geography exams:

- To 'interpret' usually means suggesting reasons for something, such as a trend (in a chart) or a pattern (on a map).
- To 'analyse' something means to break an idea or proposition apart and look at it in a detailed or structured way. For example, if you are asked to 'explain why climate is an important influence on water cycle flows', you might want to explain (i) why rainfall and therefore water flows vary from place to place, and also (ii) why climatic variations lead to variations in vegetation type, interception cover and thus water cycle flows in different places. In other words, by showing how climate has both *direct* and *indirect* influence over water cycle flows, you are analysing (and not just describing) its importance, which the question requires you to do.

- To 'evaluate' means to make a judgement about something, such as the *overall success* of a management strategy, or the *relative importance* of the way different factors and processes have shaped a physical or human environment. Evaluation is usually assessed through essay writing. You can demonstrate your ability to perform this skill *either* by writing a conclusion *or* on a more *ongoing* basis throughout your essay (for example, at the end of each paragraph you can make an evaluative comment or 'mini-conclusion'). Chapter 4 explores what a 'proper' evaluation looks like.

> **! Common pitfall**
>
> Examiners look at the *quality not quantity* of your writing when marking an AO2-targeted essay. Too many students wrongly think that the best approach to essay writing is to 'write all I know'. But more descriptive points do not necessarily mean more marks. Chapter 4 takes an in-depth look at how to carry out a proper evaluation.

Assessment objective 3 (AO3)

Use a variety of relevant quantitative, qualitative and fieldwork skills to: investigate geographical issues; interpret, analyse and evaluate data and evidence; construct arguments and draw conclusions.

AO3 has a diverse character:

- Exam boards differ in their approach to AO3. Chapter 6 guides you through the approach your own board takes with regard to AO3 in the examination papers.
- The independent investigation is viewed widely (though not universally) as being one long 'AO3 task'.
- AO3-targeted exam questions range from the very easiest kinds of 1-mark mathematical or graphical task to complex and demanding extended writing (notably so in the case of the Edexcel course). Chapter 2 focuses on the diverse AO3 skills-based questions and tasks that feature in examination papers, with attention paid to both quantitative and qualitative data.

The difference between...

Quantitative data	Qualitative data
Quantitative data consist of numerical and frequency data, which can be tabulated, tested statistically or converted into charts and graphs.	Qualitative data are non-numerical observations and descriptions of phenomena. They include in-depth interviews with people (which may read like excerpts from a newspaper or autobiography), photographs, novels, poems, paintings and even music.

Synopticity

In addition to the three main AOs, some of your marks are awarded for 'synopticity' at particular points in your exams. This is a requirement for all A-level courses, not just geography. Instead of focusing on one isolated topic, you are expected to draw together information and ideas from across your specification in order to make connections between different 'domains' of knowledge, especially links between people and the environment (i.e. connections across human geography *and* physical geography).

- Some exams achieve synopticity by presenting students with a range of resources. For instance, the WJEC and Eduqas courses show students four maps or other types of figure as the stimulus for a synoptic essay. (In one exam paper, maps showing different kinds of global risk — earthquakes, tsunamis, sea-level rise, terror attacks — served as a 'springboard' for the essay question: 'Assess the severity of the different risks faced by world cities'.) Edexcel goes a step further by producing a large resource booklet accompanying the third and final exam paper.

- Some exams achieve synopticity using a standalone essay. For example, an AQA exam paper included the essay: 'With reference to a tropical rainforest, evaluate the role of governance in environmental management'. The aim of this question was to get students to apply and evaluate ideas drawn from two different parts of the A-level course (ecosystems *and* global governance). Similarly, an OCR exam paper uses standalone synoptic questions to assess the 'geographical debates' part of the course. An example of this is the 'mini essay': 'Examine how the risks from tectonic hazards affect place-making processes'.

Exam tip

Throughout your course, take careful note of **synoptic themes** whenever they emerge in teaching and learning. Examples of synoptic themes could include the impact of urban growth on water cycle flows, or the role of tectonic hazards in creating refugee movements. Whenever you finish a topic, make a note of any synoptic themes that have emerged (perhaps keep a section of your folder or exercise book reserved for this). Synoptic exam questions are worth plenty of marks, so you need to be well prepared for them.

Using the specialised concepts

An important feature of the A-level Geography courses for England and Wales is the requirement for all learners to be familiar with a list of specialised concepts:

- **Adaptation** An ability to respond to changing events or to thrive in a particular environment.
- **Causality** The relationship between cause and effect.
- **Equilibrium** A broad state of balance between inputs and outputs in a system.
- **Feedback** The way in which changes in a physical or human system become accelerated (positive feedback), or are cancelled out (negative feedback), by internal system processes.
- **Globalisation** The process by which the world is becoming increasingly interconnected.
- **Identity** Real or perceived defining characteristics of places, objects or societies.
- **Inequality** Social and economic disparities between and within different places (and the individuals and societies who live there).
- **Interdependence** Relations of mutual dependence.
- **Mitigation** The reduction of something that is having a negative effect.
- **Representation** How somewhere, something or someone is formally and informally portrayed.
- **Resilience** The ability of an object or a population to adapt to changes.
- **Risk** The possibility of a negative outcome resulting from a process or decision.
- **Sustainability** Development that meets the needs of present and future generations.
- **System** A set of interrelated objects and/or organisms. A system's components are linked together by flows of energy and/or materials.
- **Threshold** A critical limit or level in a system or environment that, if crossed, can lead to massive and irreversible change.

Familiarity with these concepts supports knowledge and understanding of all the topics included in each exam board's course. Think of these ideas as the 'synoptic glue' which helps us make connections between what may appear completely unrelated topics, such as the carbon cycle and changing places. In this case the concept of (positive) feedback is equally important for both of these topics:

- Methane release from melting permafrost is accelerating (because methane is a greenhouse gas, which contributes to further temperature rises).
- The closure of one company in an area can lead to further supply chain closures and the out-migration of people, triggering even more business failures in 'ever decreasing circles'.

There are countless more ways of using the specialised concepts to make links between geography topics.

Activity

Work out how many of the specialised topics feature in each geography topic you study. During your 2-year course, keep notes on where the concepts appear in your teaching and learning.

Assessment levels

When your teacher or an actual examiner marks your longer answers, the first thing they will do is try to place it in a 'level'. Each type of question has different levels, and approaches vary slightly among exam boards. Typically:

- short-answer tasks (worth between, say, 5 and 10 marks) have three levels (see Chapter 3)
- evaluative essays (worth between 15 and 20 marks) have four levels (see Chapter 4)

Each level will have a description attached to it. Lower levels tend to include words like 'limited' or 'descriptive', and higher levels include words like 'accurate' and 'balanced'. Knowing these descriptors will help you understand what you are aiming to achieve. They can all be found in your exam board's sample assessment materials (SAMs) and past examination paper mark schemes. The table below offers general guidance for levels-based marking.

Table 2 Levels-based marking for A-level Geography

Highest band	Answers to the short-answer tasks are clear and factually accurate, displaying good knowledge and understanding, supported by developed examples, sketches and diagrams. Descriptions are clear.
	Answers to the evaluative essays are well written and argued so that the command word (e.g. 'discuss' or 'assess') has been followed. Knowledge is very detailed, accurate and well supported by examples, and issues are fully understood.
Middle band(s)	Answers to the short-answer tasks are often unbalanced and partial responses, which may be unstructured and make points in a random order. Knowledge is present but not always factually accurate or completely understood.
	Answers to the evaluative essays demonstrate some understanding but not of all the points, and the examples are mostly accurate and rather sketchy. Diagrams and statistical work may be less complete. The command word is interpreted wrongly to mean 'description' rather than discussion or evaluation.
Lowest band	Answers to the short-answer tasks have very limited and possibly fragmented factual knowledge. There might be no valid examples or just a single example named but not developed.
	Answers to the evaluative essays may be a set of unrelated, undeveloped ideas, possibly only in note form, and rather hit and miss in their relevance to the question. The command word might be ignored.

1 Geographical reading and note-taking

Learning objectives

- To understand the usefulness of different ways of reading and learning
- To explore different ways of note-taking and find a personal style that works well
- To develop good reading habits, including compiling a bibliography and identifying biased or fake news
- To understand how good reading habits can help you achieve a high exam grade

Study skills

Reading and note-taking

Reading and note-taking are your two core study skills. Doing both in a time-effective way is not as intuitive or as easy as some people think. In particular, it is important to read and note-take in ways that *help build up your other geographic capabilities* (alongside the acquisition of new knowledge and understanding that reading brings). This is achieved via a range of **active-reading** strategies.

The difference between...

Passive reading	Active reading
Passive reading refers to a way of reading in which the reader is not deeply engaged with the text at all. It is possible to see and understand every word that appears in a book and yet be not much the wiser for having done so. You may recall fragments of information after finishing the book, but your reading has not helped develop your broader geographic capabilities, such as thinking critically about: • the truthfulness or accuracy of statements • the degree to which generalisations can be made on the basis of evidence you have read • possible connections that can be established between what you are reading and other things you already know about People who read passively often use multi-coloured highlighter pens excessively; it may be that the mechanical act of colouring in too many words is a distraction that stops them engaging more deeply with the text.	Active reading describes reading in which the reader not only studies the words on the page but *additionally processes the information and asks themselves extension questions*. Your brain is carrying out two different tasks, rather like a phone that is running two apps simultaneously. Geographers are particularly interested in synoptic links between different topics; active reading can involve asking geographical questions about the synoptic links between what we are reading and topics we have studied previously. For instance, you might read an article about a city's growth that mentions in passing that a river runs through it. The writer does not tell you how urban growth has affected this river's hydrograph. But geographers who are reading *actively* may find themselves wondering what the impact has been on the river's hydrograph. Next time you read a geography textbook, set a target: devise one (possible synoptic) extension question per page based upon what you have read.

Active reading at A-level also means being aware of whether a writer is merely *describing* information or is additionally *evaluating* it. Descriptive writing conveys factual information in an uncritical way, for example the sentence: 'Clapham is a suburb of London whose population grew rapidly during the late 1800s because of the expansion of the railway network.' Evaluative writing goes a step further in one of several possible ways:

- The writer may offer a *view* or *judgement* about the *significance* or *role* of something. For example: 'The growth of the railway network is the most important reason why some parts of London's suburbs, such as Clapham, prospered more than others in the late 1800s.'
- Or the writer may offer a *nuanced* evaluation that explicitly acknowledges the *partial* or *tentative* nature of any causal explanation or conclusion. For example: 'While the growth of the railway network was an important reason for Clapham's growth, other factors, such as relief and drainage, undoubtedly played an important role too.'

In summary, we can read a text in two different ways: firstly, as a source of facts and data; secondly, as a series of viewpoints and arguments. Active reading that 'pulls apart' an author's writing in this way is sometimes called **deconstruction** of a text. Chapter 4 (page 58) shows how a newspaper article about urban rebranding can be deconstructed as part of the preparation process for writing an evaluative essay.

Skimming and scanning

Along with active reading (sometimes called critical reading) there are two other useful reading strategies that you can adopt from time to time. These are both 'fast' reading techniques that are used either as a first stage of engaging with a text (prior to active reading of it) or as the main way of dealing with a lengthy text that does not need thinking about carefully and is being used simply as a source of supporting facts (such as details of a TNC's supply chain, or key facts about a particular coastal landscape).

- **Skim-reading** (or **skimming**) means quickly glancing at all of the pages of a text in order to gain a general idea of what theories or case studies are included. Skim-reading may reveal that a text is not particularly useful and deals with a slightly different topic from what you had at first imagined (and can therefore be discarded). Skim-reading can also help you navigate quickly towards those paragraphs or diagrams in an article or book that have the greatest value for the task you are carrying out (such as researching case studies for your essay). There is no point devoting large amounts of time to reading a whole book or long article that subsequently turns out to have marginal or tangential relevance to your research — *always skim first*.
- **Scan-reading** (or **scanning**) means looking for particular words or phrases within an article or book chapter, or checking to see how it has been structured using different sub-headings. You might use this technique when searching for a key word definition or place name to add to an essay. You can also scan through the contents or index pages of a book to find specific words or examples to use.

Reading and note-taking at university is even more demanding than at A-level; developing a good strategy for carrying out these tasks while you are still at school or college will pay dividends in the long run.

Take it further

You can find out more about different reading and note-taking strategies using the Open University's online resources: www2.open.ac.uk/students/skillsforstudy/use-an-efficient-approach.php

Along with a reading strategy, you need a good note-taking strategy. Parallels can be drawn with the art of reading insofar as 'passive' note-taking consists of more or less copying out word for word whatever it is that you are reading. As a result, note-taking turns into a mechanical chore. Attention gets paid to making sure all the words have been copied out letter for letter instead of *thinking about what the words mean and condensing the information into a more user-friendly format*. Active note-taking strategies include the following:

- **Produce a concise summary of what you have been reading, or a list of key points.** You may be familiar with the magazine *Geography Review*: each article is approximately 1500 words long and is followed by a 100-word 'key points' box. If you are reading an article or book chapter that does not have a key points box, why not try to produce one yourself as part of your note-taking? What key messages or stories is the text telling you?
- **Try to visualise graphically what you are reading.** Much of the A-level Geography literature you study will tell you about the causes or consequences of something (urbanisation, flooding, population growth, a volcanic eruption, urban rebranding, etc.). Therefore, your notes on a topic could simply consist of a spider diagram with detailed annotations, or some other type of 'sequential' illustration. Diagrams can leave a mark on memory better than words alone sometimes do. This is called 'visual learning' and it is an effective way of taking notes actively rather than passively (page 12), as well as revising. Diagrams do not need to be hand drawn: Microsoft Word 'SmartArt' is relatively easy to use. Figure 1.1 shows a simple flow diagram to summarise an 860-word article that appeared in the *Financial Times* newspaper. The article's focus was the acceleration of out-migration from Puerto Rica following devastation caused by Hurricane Maria in 2017. It analysed how long-term economic problems were now being amplified by the hurricane's impacts.
- **Collaborate with a 'study buddy'.** Working together, you can discuss an article or textbook chapter you have been asked to read and make notes on for homework. Between yourselves, agree on what you think the most important ideas are and produce a summary of these points. Be sure to think critically about any ideas or questions that you viewed as very important but which your study buddy saw as being less important (or vice versa!) — who was right, and why?

Exam tip

If you adopt a note-taking strategy that condenses arguments or ideas into key points or a short summary, be careful to include a few key facts (such as data or names) to help back up any arguments.

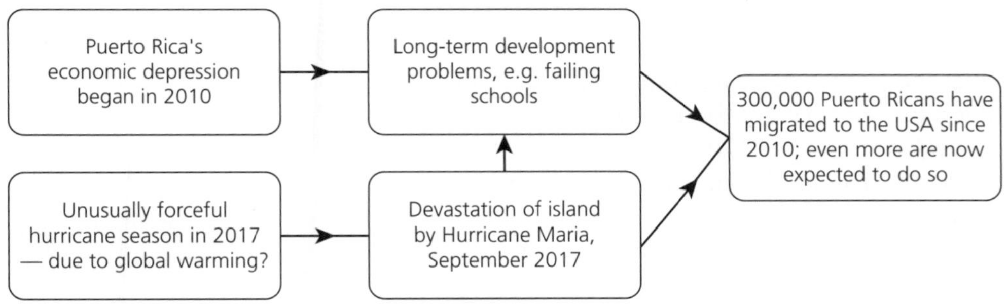

Figure 1.1 Flow diagram providing a summary of a newspaper article ('Puerto Ricans consider departure', *Financial Times*, 01.10.2017)

Activity

Find a 1000-word newspaper article (either online or in print) that relates to a topic you are currently studying. Try to produce a visual summary of it (either based on Figure 1.1 or using your own ideas).

Wider reading

Core textbooks provide supporting information for each topic you study. Aspiring A/A* students should be reading far more widely, however (especially those who are planning to read geography at university). Table 1.1 suggests a range of possible types of literature you could make use of, along with an idea of what you might gain from doing so. Note that while some of these sources are free to consult online, others might require you to pay a subscription (albeit at a reduced rate if you provide evidence that you are a student).

Table 1.1 Opportunites for wider reading

Newspapers (print copy or website)	Quality newspapers such as the *Guardian* and *Financial Times* provide up-to-date information on the latest geopolitical, economic and environmental events. Home (or domestic) news stories may tell you how processes such as migration and investment are affecting particular places in the UK; world news updates give you the latest developments in climate change governance or global trade flows. Remember that course textbooks go out of date quickly; GDP figures or oil prices sometimes vary greatly from year to year. Teachers and examiners may be impressed to see contemporary figures from recent print media in your answers.
Periodicals (print copy or website)	*National Geographic* is a widely used periodical that (i) provides up-to-date information and (ii) demonstrates well the art of evaluative writing. *National Geographic* investigative journalists spend many months researching their stories and often develop strong arguments, using the facts that they have uncovered. Weekly publications such as the *Economist* and *New Statesman* provide excellent support for human and environmental geography. Their 'opinion pieces' and longer articles blend fact and argument in ways that you can learn a lot from.
Online academic papers	Occasionally you may try to read an article online written by a professional geographer for a target audience of other geographers. Articles appearing in journals such as *Transactions of the Institute of British Geographers* are aimed at a mature audience of geography lecturers and their undergraduates. Do not be surprised if a lot of the language and terminology is not always easy to understand. Skim-read academic journal articles first in order to check that they are (i) relevant to your research and (ii) relatively easy to read.
Government, business and non-governmental organisation (NGO) information	The websites of local councils, government agencies (such as the Environment Agency), transnational corporations (Shell, Apple), consultancy firms (McKinsey) and NGOs (Amnesty International, Greenpeace) can be carefully 'mined' for factual information as part of your studies. These websites provide you with free-to-download PDF files, which you can store on your computer and add virtual sticky notes to highlight particularly useful facts or passages.

Application to the exam

Developing good reading skills and habits does not only make sense by helping you to study more effectively, there are additional virtuous impacts on *how you perform in your assessments and the final grade you receive*. In particular, the marks you receive may be influenced by whether you can:

- include a high-quality bibliography in your independent investigation
- write critically about the reliability of information and possible bias in secondary data
- spell technical terms and place names accurately

Producing a bibliography

It is very important to keep an accurate record of the literature you review when producing your independent investigation (see also Chapter 5). The AQA, OCR and Edexcel specifications explicitly mention the importance of using a bibliography system.

- Make a note of your sources and use the Harvard referencing system when citing (referencing) them. Each academic reference should contain, in sequential order: (1) Name of author(s), (2) Year of publication, (3) Title, (4) Publisher and (5) Pages used (if applicable).
- The approach differs slightly if you cite a newspaper or internet source. For further guidance, look at the examples in Table 1.2.

Table 1.2 Citing references

Newspaper article	*Financial Times* (2017) Puerto Ricans consider departure after Hurricane Maria, 2 October
Journal article	Cloke P, Milbourne P and Thomas C (1997) Living lives in different ways? Deprivation, marginalisation and changing lifestyles in rural England *Transactions of the Institute of British Geographers* 22(2), 210–230
Online technical paper	Environment Agency (2001) *Lesson learned: autumn 2000 floods* (available online at www.gov.uk/government/uploads/system/uploads/attachment_data/file/292917/geho0301bmxo-e-e.pdf; last accessed 10.10.2017)
Book	Williams R (1973) *The Country and the City*, Chatto and Windus, London Massey D (1993) 'Politics and space/time' in Keith M and Pile S (eds) *Place and the Politics of Identity*, Routledge, London, 141–161

Bias and fake news

An important aspect of active reading is thinking critically about whether what you are reading is accurate and trustworthy. Both your independent investigation and exam papers provide opportunities for you to demonstrate understanding of bias and so-called 'fake news'.

Biased reporting does not set out to tell deliberate lies, but it deliberately omits 'inconvenient' facts or evidence in order to strengthen a partisan (particular) viewpoint.

- Complaints about media bias have existed for as long as newspapers have been in circulation. In the days before social media, voters for different political parties tended to purchase those newspapers whose biased views mirrored their own most closely.
- Traditionally, the *Sun* and *Daily Telegraph* have reported political stories in a way that appeals to more right-wing voters, while

Exam tip

An exam question could ask you to write about the 'usefulness' or possible 'limitations' of a quantitative or qualitative data source. Possible bias in the data (especially if they include words or images) may be a theme you can write about.

the *Guardian* and *Mirror* journalists cover the same events with more liberal readers in mind.

- Bias can often be seen in reporting about immigration, for instance. One newspaper may inform its readers about the potential security threat posed by a minority of refugees, while neglecting to mention a new report that shows how immigration brings economic benefits. Another newspaper may do the exact opposite.

So-called 'fake news' goes a step further in its reporting. One article widely shared on Facebook in December 2016 said that Syrian terrorists posing as refugees had attacked New York. No such attack had taken place. 'Fake news' means a *completely inaccurate* or made-up tale that has been written and presented (usually online) in a way that makes it appear to be an authentic (and supposedly truthful) mainstream news story. Some fake news reports are politically motivated, while others are written purely for financial gain (Figure 1.2).

Figure 1.2 Examples of fabricated 'fake news' stories

Spotting biased or 'fake' secondary data

It is much easier to collect secondary data than it used to be. In the 1990s most geography students still had to physically visit libraries when searching for background information to support their independent investigations. Research can now be carried out online. However, the fake news controversy shows how important it is to keep a record of who has authored any information you use (this is called 'sourcing' your data). The websites you visit may differ greatly in terms of their credibility and trustworthiness:

- **Formal** websites are produced by a government, university, large company or news service. The information is likely to have been checked for accuracy by many different people.
- **Informal** websites are sometimes produced by a single person working alone, possible anonymously, and may contain accidently or deliberately inaccurate material.
- **Social media** and 'creative commons' are online spaces where anyone can say whatever they like about anything, subject to certain rules and norms. Bias and lies may be widely tolerated.

It is good practice to always question the validity and reliability of information found on the internet. After all, anyone with the right technical skills can build a professional-looking website. Figure 1.3 shows important questions you can ask when carrying out your own research. Three particularly important rules are:

1. Don't assume that the credibility of a source is shown by how high a story appears in Google search results. Fake news is often very popular, and is ranked highly accordingly.
2. Be aware of your own biases — we tend to fall for fake news stories more easily when the story corresponds with our own worldview.
3. Don't think you will always be able to recognise invalid or fake data and news. These days misleading information is sometimes presented in such a professional way online that almost everyone will be fooled.

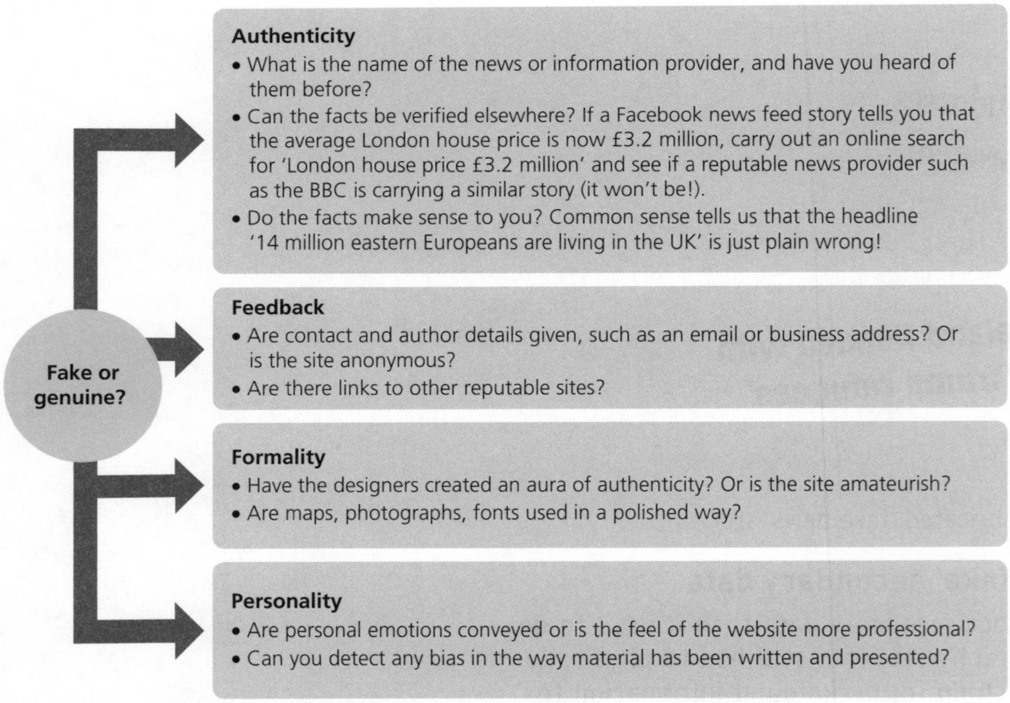

Fake or genuine?

Authenticity
- What is the name of the news or information provider, and have you heard of them before?
- Can the facts be verified elsewhere? If a Facebook news feed story tells you that the average London house price is now £3.2 million, carry out an online search for 'London house price £3.2 million' and see if a reputable news provider such as the BBC is carrying a similar story (it won't be!).
- Do the facts make sense to you? Common sense tells us that the headline '14 million eastern Europeans are living in the UK' is just plain wrong!

Feedback
- Are contact and author details given, such as an email or business address? Or is the site anonymous?
- Are there links to other reputable sites?

Formality
- Have the designers created an aura of authenticity? Or is the site amateurish?
- Are maps, photographs, fonts used in a polished way?

Personality
- Are personal emotions conveyed or is the feel of the website more professional?
- Can you detect any bias in the way material has been written and presented?

Figure 1.3 Fake or genuine? Ways of assessing the trustworthiness of online data and news sources

Activity

You can find many lies about Oxford online that were created by the *Oxford Mail* newspaper as a 'prank' to fool tourists. Think about what the implications of this are for an A-level student who is researching Oxford as a 'Changing places' case study. You can also practise your research skills by trying to find details online of the fake news 'Tweet-a-thon' (as reported by the *Oxford Mail* in 2014).

Grammar and spelling

No specific marks are set aside for spelling and grammar at A-level. However, poor spelling and grammar can adversely affect the grade you receive.

Grammar, clarity and coherence

In their marking criteria for the independent investigation and question papers, the exam boards reserve the highest marks for answers that analyse and evaluate geographical information in 'clear' and 'coherent' ways. Poor grammar can damage the clarity and coherence of your arguments. Therefore, you should try to take note of how the books and articles you read are phrased and argued (and not just the facts and ideas they provide).

! Common pitfalls

By improving your grammar you can avoid the following common pitfalls, all of which can affect the mark your work is awarded (because of the knock-on effect for clarity and coherence of argument):

- Switching between past and present tenses when recalling a case study.
- Phrasing that muddles cause with effect, or fails to clearly articulate the effect (explanation may be hindered by the limited use of connectives, such as *so*).
- Lack of clarity as to who the subject of the sentence is (or repetitive use of the word 'they' when describing the actions of different groups of people in a case study).
- Confusion over the correct use of *it's* and *its*.
- Confusion over the use of *effect* and *affect* (*effect* is usually a noun, meaning 'result', whereas *affect* is a verb, meaning 'to influence' — human activity can *affect* the physical environment; the physical environment has numerous *effects* on human activities).
- Misuse of possessive apostrophes (for example, when describing the actions of local stakeholders in a case study).

Spelling it out

When you are reading and taking notes about unfamiliar new ideas or places, make sure you spell out important words carefully. Misspelling of important words in the exam may result in you gaining a lower mark than you deserve. This is because credit may not be awarded if the examiner cannot work out what you have written.

- Some specialist terms are commonly misspelt — even by potentially high-scoring A-level students. These include: hydraulic (radius), (trade) bloc, and desert. Some 'everyday' misspelt geography words that are important elements of a geographical vocabulary include: environment, vegetation, development and business.

Exam tip

In the exam, keep your sentences short and punchy, as newspaper journalists do, in order to tighten up the grammatical structure of each sentence. To prepare you for the exam, pay attention to structure during your reading of newspaper articles related to your study topics — newspaper journalists are trained to communicate information and argue viewpoints in clear and coherent ways.

→ Some place names that are commonly misspelt include Mumbai, Antarctic and Cairo.

It may be helpful to compile a glossary of technical terms and example place names for each of your A-level Geography topics. Writing out key words helps them become more familiar and reduces the chances of misspelling them in the exam.

You should know

- **Active reading helps you develop a wide range of vital skills in addition to learning new information.**
- **It is a good idea to skim-read an article or chapter for relevance before you begin reading it properly.**
- **Effective note-taking involves summarising material, not copying it out word for word.**
- **Note-taking can be done in interesting ways, including visualising information (find a way that works well for you).**
- **It is important to think critically about whether the reading sources you use are reliable and trustworthy (be prepared to write about this important issue in your exams and independent investigation).**
- **To access the top marks in the exam and independent investigation, clear and coherent writing — including accurately spelt key words — is often essential.**

2 Geographical data analysis skills

Learning objectives

- To carry out data skills-based tasks accurately and write about statistics in a clear and coherent way
- To analyse different kinds of patterns and trends effectively
- To find meaning and significance in geographical data and communicate this clearly
- To think critically about different data presentation and analysis techniques

Under assessment objective 3 (AO3) for A-level Geography, you must demonstrate that you can *use a variety of relevant quantitative, qualitative and fieldwork skills to: investigate geographical questions and issues; interpret, analyse and evaluate data and evidence; construct arguments and draw conclusions.*

These AO3 capabilities are assessed through the completion of an independent investigation (see Chapter 5) and also in your examination papers. While the precise targeting approach for AO3 varies for the different specifications (see Chapter 6), common types of examination question include:

- **Graphical and mathematical tasks**, such as plotting data on a map or chart, or completing a simple mathematical calculation such as the mean or interquartile range for a data set. Often, only 1 or 2 marks will be available for these tasks.
- **Short analytical questions** asking you to 'describe', 'analyse' or 'compare' charts, maps, graphs, diagrams and photographs. Typically, between 3 and 6 marks are available for such tasks (depending on which specification you are following).
- **Short evaluative questions** asking you to comment on possible strengths or weaknesses in the way information has been presented.
- **Extended writing**, which incorporates data analysis. This may take the form of an evaluative essay (see Chapter 4), which requires you to make use of information in a resource booklet that has been provided as part of the examination paper. You are expected to make selective use of this information — in conjunction with your own ideas and case studies — in order to build an argument or argue a viewpoint. The demands of these extended-writing questions are not covered in this chapter. Instead, they are dealt with in Chapter 6, which looks at the varied approaches of each different specification in turn.

AO3 questions are often found at the start of an examination paper (or at the start of each different section of an examination paper). There is often a perception that these 'starter' questions are relatively easy, perhaps because they require you to demonstrate your competence in performing a mechanical task rather than writing in detail about complex ideas and theories. It is certainly true that most examination papers are designed with an inbuilt 'incline of difficulty'. This means that the demand of questions is 'ramped up' as you progress through each section of the examination. Generally, the longest and most demanding essay-style questions are positioned last.

There may therefore be some truth to the claim that AO3 'starter' questions are, by design, not always designed as especially challenging tasks. Indeed, reports on past A-level Geography examination papers have sometimes included comments such as: 'Most candidates had no problem completing the graph in Question (a)(i) accurately'.

Exam tip

At the outset, it is important to recognise that **communication (literacy** and writing) and **numeracy** skills — and not just your geographical knowledge — provide essential foundations for carrying out AO3 tasks in A-level Geography examinations. If you under-performed at GCSE English or Mathematics, then you may need to take every opportunity to practise answering AO3-targeted examination questions in A-level Geography.

However, it is important not to be *too* relaxed about completing analytical tasks:

- Some geography students who are very good at mathematics may complete statistical tasks included in the exam in a rushed and overconfident way, resulting in careless errors and a lower grade than expected.
- Other students continue to underestimate the importance of communication skills. In order to succeed fully at answering questions that ask you to 'describe' or 'analyse' a graph, you must be literate as well as numerate. Failure to use the right technical words and phrases when describing a complex pattern or trend may result in supposedly 'easy' marks being lost.

Study skills

During your 2 years of study, you are expected to develop a range of geographical skills and become proficient in carrying out a range of 'mechanical' techniques and procedures. These are shown in Table 2.1 and are common to all A-level Geography specifications. Your school or college may take one of two approaches:

- It may instruct you how to carry out all of the techniques and procedures listed in the specification in special 'standalone' lessons devoted to geographical skills.
- It may 'embed' learning about all of the different geographical skills at intervals in regular topic-based lessons. For example, during one lesson on coastal landscapes you might be asked to calculate the mean, mode and median value for a sample of sediment as part of your landscape studies; your teacher would then expect you to be able to use this procedure, if asked to, in an examination question about 'Changing places'.

Common pitfall

If a particular technique has been 'embedded' by your teacher in a lesson that you miss due to illness, it is vital that you master this skill in your own time before the final examination. You cannot afford to lose potentially easy marks in an examination because you have a 'skills gap'. Make sure that you are familiar with *all* of the skills listed in your specification.

Table 2.1 Geographical skills common to all specifications

Quantitative data	Qualitative data
Students must demonstrate the following skills specific to quantitative data: • Understand what makes data geographical and the geospatial technologies (e.g. GIS) that are used to collect, analyse and present geographical data. • Demonstrate an ability to collect and to use digital, geo-located data, and to understand a range of approaches to the use and analysis of such data. • Understand the purposes and differences between the following and be able to use them in appropriate contexts: – descriptive statistics of central tendency and dispersion – descriptive measures of difference and association, inferential statistics and the foundations of relational statistics, including (but not limited to) measures of correlation and lines of best fit on a scatter plot – measurement, measurement errors, and sampling	Students must demonstrate the following skills specific to qualitative data: • Use and understand a mixture of methodological approaches, including interviews. • Interpret and evaluate a range of source materials, including textual and visual sources. • Understand the opportunities and limitations of qualitative techniques such as coding and sampling, and appreciate how they actively create particular geographical representations. • Understand the ethical and socio-political implications of collecting, studying and representing geographical data about human communities.

Plotting data

During your A-level course you may be asked to plot data on a graph as part of a homework assignment. In theory, this seems like a skill that all A-level students should be able to carry out without losing marks. However, there are a few pitfalls that aspiring A-grade students must avoid. It is a shame when someone falls a mark or two short of gaining a really good grade in a piece of work because credit has been lost due to a careless mistake when carrying out a simple task. The moral is: don't be overconfident! When plotting a graph always take care to:

- use a pencil — that way, if you do make a mistake, you will not have made a terrible mess of the graph
- pay attention to any symbols that are used and copy them as best you can. If the points on a scatter graph are presented as diamond shapes then do not draw crosses when you are asked to plot additional data. Draw diamonds.
- pay attention to whether or not the graph uses a logarithmic scale (Figure 2.1), and make sure that you are familiar with the conventions of logarithmic graph paper

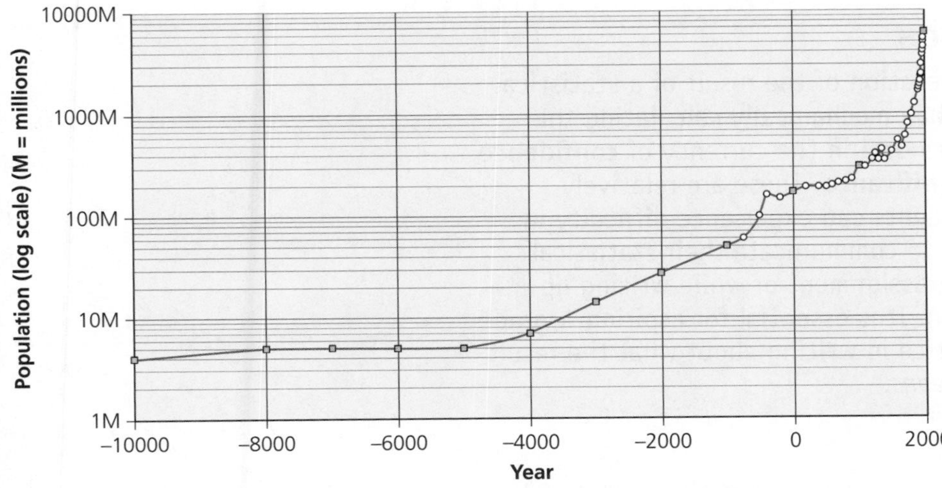

Figure 2.1 A semi-logarithmic graph showing world population growth during the last 12,000 years

Activity

Make sure that you can plot data accurately on a logarithmic graph. Plot values of (–4000, 20M) and (0, 7000M) on Figure 2.1. Why might some people plot data inaccurately on a logarithmic graph?

Making calculations

Throughout your course, you are expected to master all of the quantitative skills shown in Table 2.1. These include relatively simple calculations, one or two of which you may be expected to carry out in your final examination. There is also an expectation that you will learn how to complete some basic statistical procedures using mathematical formulae, both as part of your independent investigation (see Chapter 5) and in the examination. Even if you found GCSE Mathematics challenging, try not to be too daunted by the mathematical demands of the A-level Geography course. Remember the following:

- Some of the calculations you will be asked to carry out really are very straightforward, such as the procedure to identify the mean, mode or median value of a data set, or to calculate the range or interquartile range for a sample of data.
- Even the more complex procedures involving the use of statistical formulae — including the Spearman test and chi-squared test — are relatively easy to carry out providing you follow mathematical conventions. You are not required in any case to memorise the formulae.

Take it further

Calculating the interquartile range requires care if an additional calculation needs to be performed because the median value and/or quartiles are not easily identifiable. This will happen, for example, if the data set contains 8, 9 or 10 values. Make sure that you know what to do. If in doubt, carry out research online (for example, the University of Surrey provides good online guidance on the median and quartiles).

Interpreting statistics

Many students find the interpretation of the result of a statistical test a more challenging task than mechanically calculating the result. This is because statistics deal in the currency of **confidence levels**, **critical values** and **significance**. These are relatively complicated ideas so some students can experience difficulty in finding *exactly* the right words to communicate their statistical findings as part of a homework assignment or while writing up their independent investigation. It is essential for aspiring A-grade students to get plenty of practice in writing about what the result of a statistical test actually *means*.

Interpreting the result of a Spearman correlation test

Let us take a look at how to interpret the result of a statistical test using the Spearman rank correlation coefficient. When this test has been completed, the result (*r*) is a number no larger than +1 and no less than –1. These two values indicate a perfect positive and perfect negative correlation, respectively. The formula used is:

$$r = 1 - \frac{6\Sigma d^2}{n^3 - n}$$

where *d* = difference in paired ranks and *n* = number of cases.

Imagine that we have collected gross domestic product (GDP) and human development index (HDI) data for 14 countries and have calculated a Spearman rank correlation coefficient of 0.69 using the formula above. The number suggests a correlation, but how strong or weak is it? How sure can we be that there really is a correlation?

You must be able to interpret this result using the **critical values** shown in Figure 2.2.

Degrees of freedom	Significance level (two-tailed) 0.05	0.02	0.01
5	1.000	1.000	
6	0.886	0.943	1.000
7	0.786	0.893	0.929
8	0.738	0.833	0.881
9	0.683	0.783	0.833
10	0.648	0.746	0.794
12	**0.591**	0.712	0.777
14	0.544	0.645	0.715
16	0.506	0.601	0.665
18	0.475	0.564	0.625
20	0.45	0.534	0.591
22	0.428	0.508	0.562
24	0.409	0.485	0.537
26	0.392	0.465	0.515
28	0.377	0.448	0.496
30	0.364	0.432	0.478

Figure 2.2 Critical values for Spearman rank correlation coefficient

Because there are 14 pairs of data, we say there are '12 degrees of freedom' in statistical jargon (this is calculated as the number of pairs of data minus 2). Figure 2.2 shows that the critical value for a 5% (or 0.05) likelihood of the correlation occurring by chance is 0.591 when there are 12 degrees of freedom. The critical value for a 1% (or 0.01) likelihood is 0.777.

But what does all of this actually mean?

- Firstly, it's important to *really* understand what's going on here. The whole point of statistical testing is to try to ascertain exactly how certain we can really be that something is true (or not). Your 'gut feeling' when looking at a collection of points on a scattergraph may be that there is not really a correlation there. However, another person might look at the same distribution and feel that there *is* a correlation. Who is right and who is wrong?

Exam tip

Make sure that you know what the different components of the formula represent and also that you can calculate the Spearman rank correlation coefficient. If you are unsure, use the Royal Geographical Society's 'Data skills online' resources at www.rgs.org/schools/teaching-resources.

- The Spearman test helps to address this uncertainty by establishing a level of confidence: once we have calculated the result and consulted the critical values table, we should be able to say we are '90% sure' or '99% sure' that we are really seeing a correlation.
- Another way of saying that you are '95% sure' of the correlation is to say 'there's a 19-in-20 chance that there really is a relationship'. Or you could say 'there's only a 1-in-20 chance that what we are seeing is not really a relationship'. Sometimes, it is worth expressing yourself in a couple of different ways like this in order to make sure that you have got your message across.

Activity

Using Figure 2.2, practise writing out interpretations of the following Spearman test results, using the guidelines given above:

- 22 pairs of data are tested and the result is +0.56.
- 12 pairs of data are tested and the result is −0.86.
- 20 pairs of data are tested and the result is +0.45.

Take it further

Reports by the Intergovernmental Panel on Climate Change (IPCC) are very carefully worded in order to attach a level of certainty to their findings. For instance, in 2007 the IPCC thought it '*very* likely' that anthropogenic GHGs were the main cause of global warming. The 2013 assessment says that it is '*extremely* likely'. Technically, this means the IPCC is now 95–100% certain (a 1-in-20 chance that the projections are wrong). The 2007 assessment was 90–100% certain (giving a higher 1-in-10 possibility of error).

Analysing relationships, patterns and trends

Being able to identify and communicate your understanding of relationships, patterns and trends are core geographical skills. The demands of A-level Geography are significantly higher than they are at GCSE in this respect. The next few pages are devoted to an exploration of some of the more demanding visual representations of relationships, patterns and trends (including maps, charts, graphs and tables) that you might expect to encounter in a typical textbook during your course. As you explore each representation, two good questions to ask are:

- What is the *story* this picture is trying to tell us? If you were writing a newspaper article based on this illustration, what would its headline be?
- How might a geographer's analysis of a chart or graph differ from one carried out by an economist or a historian? In other words, in what ways can your own visual analysis demonstrate that you are actively 'thinking like a geographer'?

Take it further

Apply the analytical techniques shown on these pages to other graphs, charts and maps in your course textbooks or to images you find online. The more you practise trying to communicate patterns and trends, the better you will become at doing so. Strive to think critically and engage actively (see Chapter 1) with the images and illustrations that accompany texts you are studying.

Activity

Go back and remind yourself of the specialised concepts for A-level Geography shown on page 10 before you read any further. Although this chapter is devoted to the analysis and description of information, rather than knowledge and understanding of geography topics, the specialised concepts can be used to support an A03 task, as you will see.

Analysing relationships

You are likely to have seen Figure 2.3 (below) before in a textbook, or something very like it. It is a standard scatter plot showing the relationship between GDP per capita (a proxy for income, and the independent variable) and life expectancy (the dependent variable).

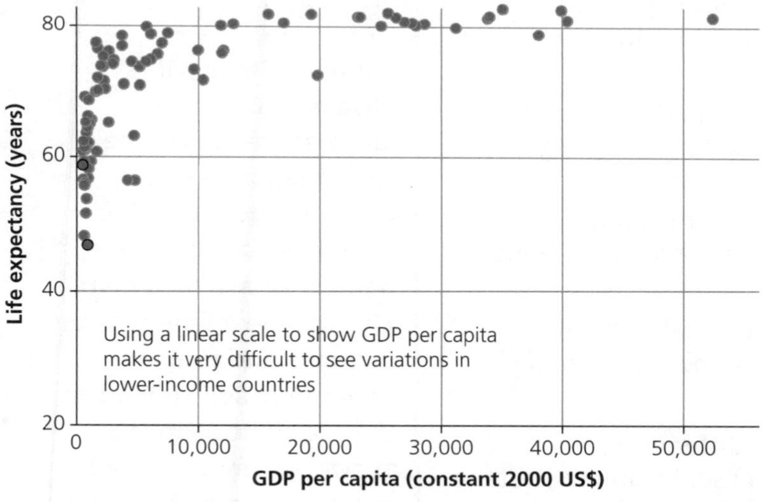

Figure 2.3 A scatter plot showing the correlation between GDP per capita and life expectancy for around 100 countries

Reproduced by permission of the *Financial Times*

At first glance, you might think that the headline story here is: 'People live longer if they have more money'. But there is an even better geographical story here, and one that a student working towards achieving an A-grade might identify: 'Having a great deal more money *does not* help you live that much longer'.

A broadly linear relationship is only seen for GDP per capita values lower than about US$5000. Thereafter, the line flattens out. This raises an interesting geographical question about correlations and *causality* (which is one of the specialised concepts on page 10). In this graph, increasing income is *not* a cause of significantly greater life expectancy beyond the threshold value of around US$10,000. And if high-income countries (US$10,000 or greater) only were shown on the graph, you would see no evidence of a correlation.

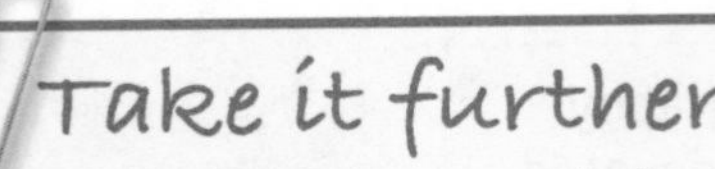

Look for other scatter graphs in your course textbooks. Think critically about the nature of any positive or negative correlations shown. Do they appear as linear or nonlinear relationships? What do the data suggest?

Common pitfall

It would be a mistake to draw a straight best fit line on Figure 2.3 because the relationship shown is **nonlinear**. A straight best fit line would imply that people live infinitely long lives in a country with infinitely large GDP per capita — a proposition which is utter nonsense. When you are working on your independent investigation (Chapter 5) think carefully about whether or not you should be attempting to add a straight line to any scatter graphs you produce.

Analysing patterns shown on maps

When describing a pattern or distribution on a map, some students rely on the use of a **mnemonic**, whereby each letter of a word serves as a reminder of an analytical point that must be made. For example, the word 'OPPAD' is a mnemonic that prompts students to:

- describe the **O**verall picture or story shown (is it even or very uneven?)
- analyse the **P**attern, if one is present (are there signs of clustering, or is a more scattered or random distribution shown?)
- identify **P**lace names worthy of special mention (for example, the distribution may be clustered around a particular city or country)
- identify **A**nomalies or map features that do not conform well with the general pattern (for example, volcanic hotspots on a world map of volcanic activity)
- include **D**ata in the answer (if the map has a scale then it is possible to estimate the length of a linear distribution of settlements, for example)

There are alternative mnemonics favoured by different teachers and schools. You may have one that you prefer to use. Whatever strategy you employ, be sure to take every opportunity to practise analysing patterns shown on maps during your course. Whenever you have completed a map analysis, take a look at what you have written and ask yourself: would somebody *who has never seen the map* be able to sketch out a crude approximation of it using what you have just written? If the answer is 'no' then you have not done the job properly.

Common pitfall

Some students will always write far too many sentences in their examination answer booklet when asked to 'describe' or 'analyse' a pattern shown on a map. This can be a time-wasting exercise and is also potentially self-defeating. Very long descriptive answers sometimes fail to score full marks because the examiner begins to feel that the student 'cannot see the wood for the trees'. Very long descriptive answers begin to resemble a *list* and the most important message — the overall picture and pattern — may become lost or obscured. As a result, full marks are not awarded.

Activity

Write a description of the pattern shown in Figure 2.4 that is no more than 70–80 words long and covers each of the points included in the OPPAD mnemonic.

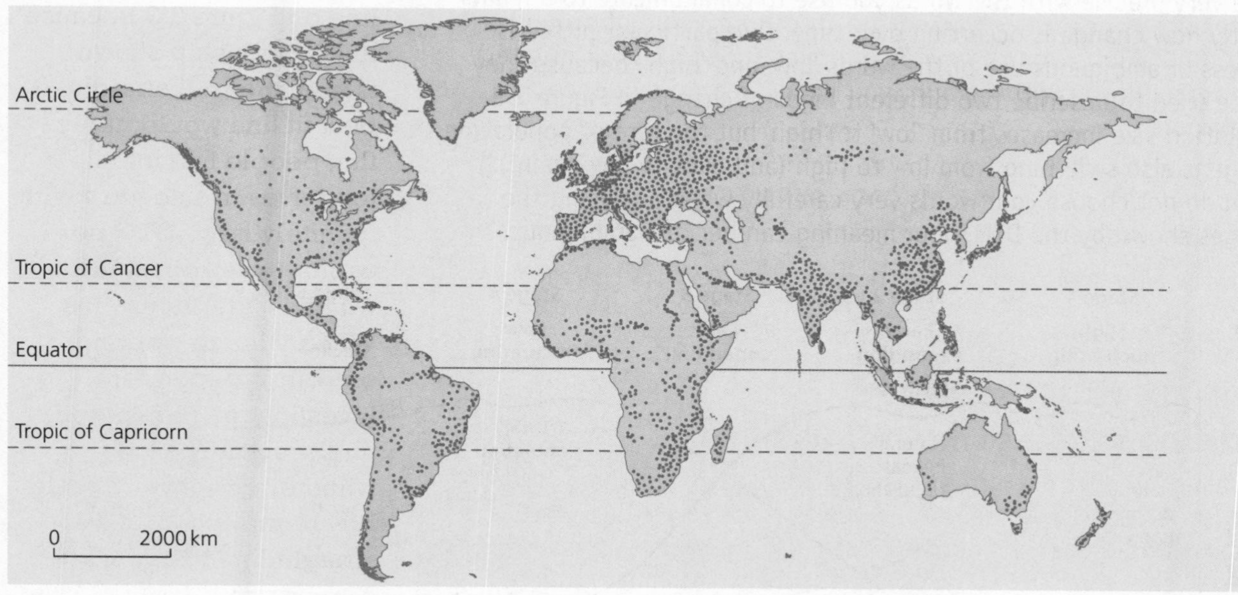

Figure 2.4 Global population distribution dot map

Activity

Your A-level examination could include a harder analytical task if a more sophisticated form of mapping is used that combines different kinds of information. For example, Figure 2.5 is a graphical representation of (1) the uneven distribution of migrants in the UK and (2) how attitudes to migration vary from place to place. A challenging practice question for you to try to answer is:

Analyse the relationship between the proportion of migrants found in different places and social attitudes towards migration in those places.

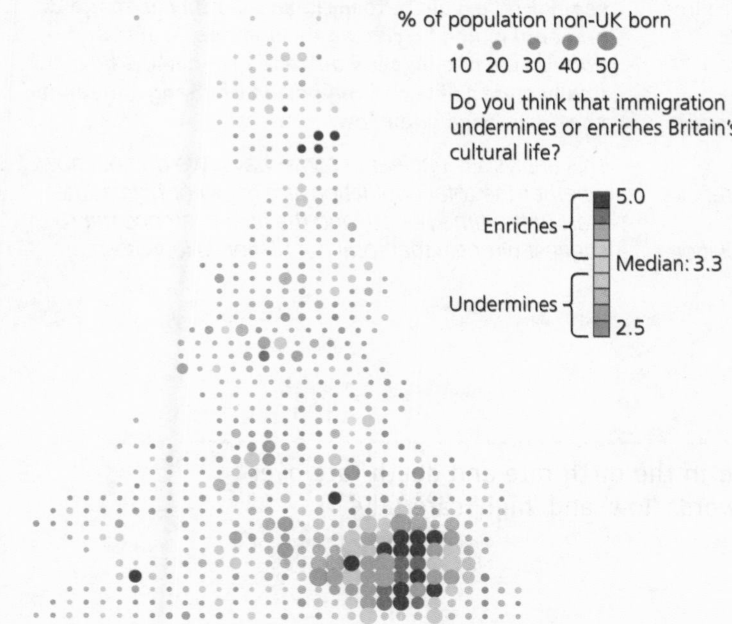

Figure 2.5 How UK attitudes towards immigration vary locally in relation to the size of local migrant populations, 2011

Analysing trends

Figure 2.6 shows the demographic transition model (DTM), which you may have encountered previously at GCSE. Analysing the trends shown in the DTM can be a lot easier said than done. This is because you need to be very precise with the words you use to communicate to a reader exactly how change is occurring over time. One particular pitfall is careless or ambiguous use of the words 'low' and 'high', because they can be used to describe two different kinds of change in Figure 2.6. Population size increases from 'low' to 'high' but the rate of population change is also switching from low to high (and back to low again!). If you do not choose your words very carefully when analysing the changes shown by the DTM, your meaning can become ambiguous.

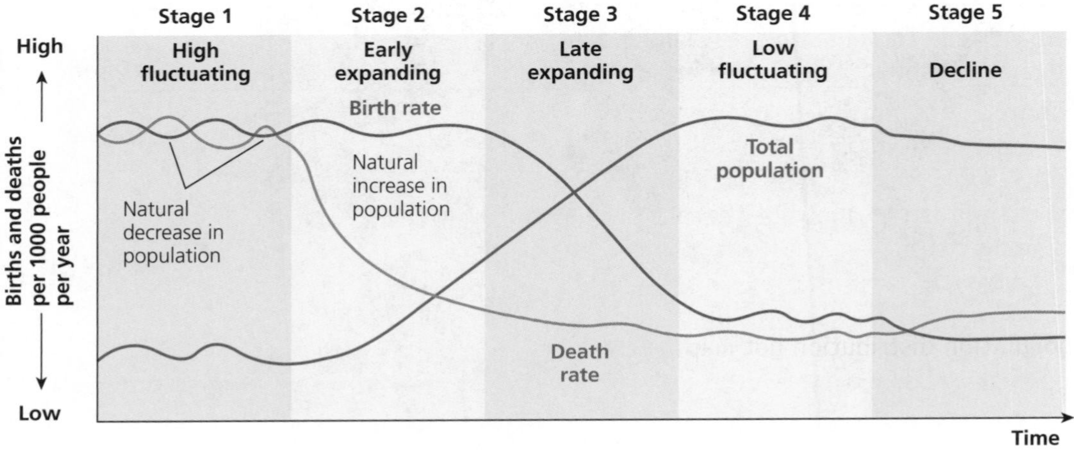

Figure 2.6 The demographic transition model

The difference between...

A-grade analysis of the total population changes over time shown in Figure 2.6	C-grade analysis of the total population changes over time shown in Figure 2.6
Total population begins at a very low level. In stage 1 there are slight increases and decreases over time prior to a rapid increase in numbers in stage 2. Population continues to rise in stage 3 but at a decreasing rate before levelling out in stage 4. Thereafter, population remains high but with some slight fluctuation in numbers. *This analysis is clear and unambiguous. The differences in the changing size of population and the rate of population change (two different aspects of population change) are distinct from one another.*	At the beginning, population rises and falls at regular intervals before rising steadily in stage 2 and continuing to rise again in stage 3. In stage 4, population has levelled out and growth has now finally ended after a long period of change, meaning that it is once again low. *This analysis is unclear in some ways. We do not know whether the total population size is low or high at the start of the time period, and the final sentence may suggest wrongly that total population is low again.*

Activity

Practise describing the changes that occur in the birth rate and death rate over time in Figure 2.6. Remember to use the words 'low' and 'high' carefully.

Ambiguous trends

Data can be political. Different people will sometimes interpret information in varying ways because of the biases and viewpoints that they hold (as a rule, people like to see their beliefs confirmed and not challenged too much). In particular, climate change data have become extremely controversial. All A-level Geography courses require that the carbon cycle is covered, which means that climate change is a topic you will certainly be required to study. Climate change data provide an excellent opportunity for you to further develop your analytical study skills. Figure 2.7 is a particularly good resource to practise writing about. Here are some possible analytical tasks to carry out:

1. Identify the maximum and minimum values. [An easy A03 task.]
2. Describe the changes shown since 2000. [A relatively easy task, provided that you have read the question carefully and do not describe all of the changes shown since 1910.]
3. Use the data in Figure 2.7 to argue for and against the proposition that temperature is increasing in the UK. [A far more challenging A03 task; no actual subject knowledge of rainfall or climate change is required but there is still a very high demand here in terms of your ability to use evidence *in order to make an argument*.]

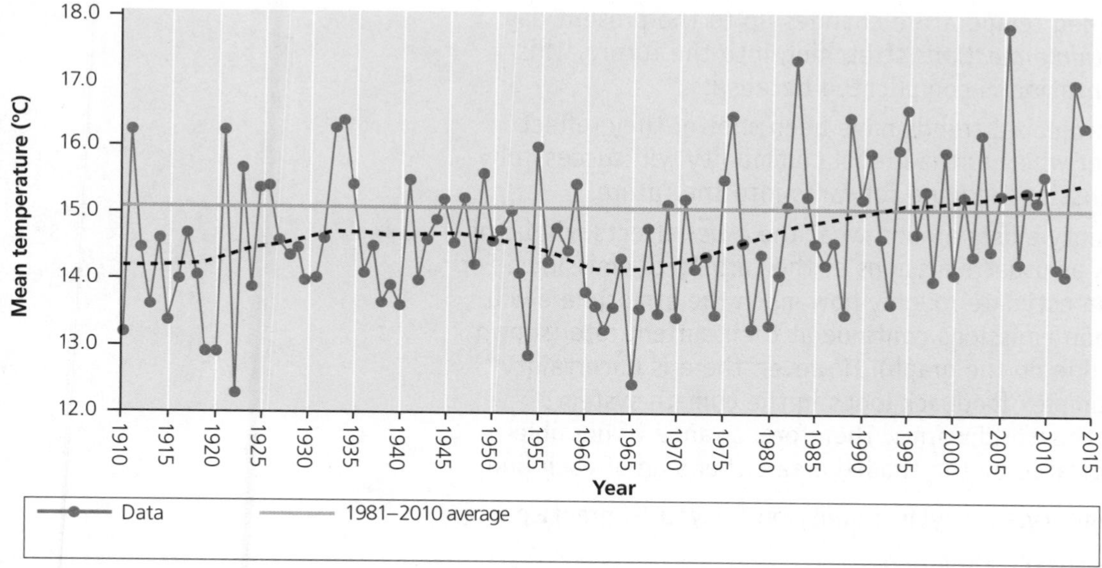

Figure 2.7 July temperature in the UK 1910–2015

Activity

Imagine you are a geography student in the year 1965 and you have been shown the data for the period 1910–65. What conclusion about changing weather patterns might you draw from the data for those years only? What does this tell us about the need to be careful when making projections based on a small sample of time-series data?

Uncertain projections

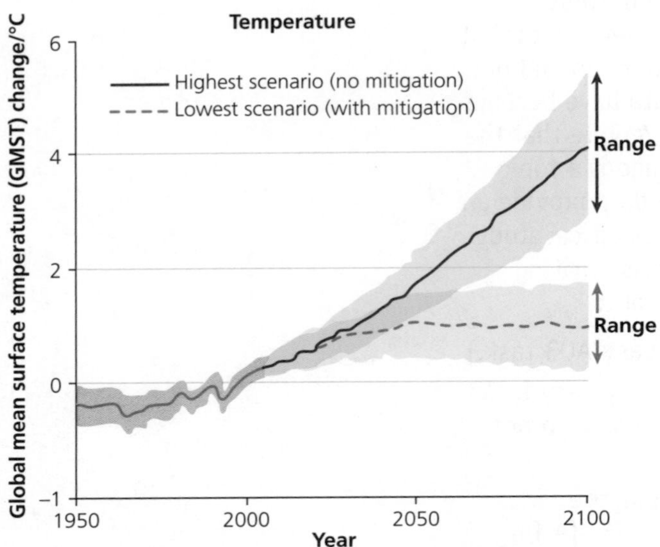

Figure 2.8 The IPCC's range of projected global temperature changes for 2100 (dependent on whether climate change mitigation occurs or not)

Figure 2.8 shows another common data presentation technique used in textbooks, academic articles and examination papers alike. This graph shows recorded temperature changes up to the present day followed by *uncertain projections* stretching into the future. This particular representation is complicated because:

- two different projected trends have been shown; they reflect uncertainty over whether the global community will successfully curb carbon emissions moving forwards into the future
- further uncertainty exists around what the *exact* effects will be of either 'business as usual' emissions or the curbing of emissions. For example, we can estimate roughly how much the atmosphere will warm by if carbon emissions continue at their current rate (shown by the steeper line on the graph). However, there is uncertainty over the way complex feedback loops in the climate system will operate as the world warms. Therefore, a range of possible outcomes is indicated by the shaded area either side of each line

A challenging exam-style analytical question for you to practise is:

Compare the two trends shown.

This is a deceptively simple question, which is potentially very hard to answer. Some teachers use **analogies** to help students find the right words to analyse complex visualisations of data such as this. Here is an analogy that can be used to help you engage with Figure 2.8:

Emma currently scores mostly grade E in geography tests. A teacher projects that if Emma works very, very hard she may gain a grade B in her final exams. However, the teacher admits that there is some uncertainty here: for example, Emma may do less well if the exam paper is very hard or contains topics that Emma missed lessons for. Or Emma may do even better and gain an A grade if the exam paper luckily includes all of Emma's favourite topics. The teacher also presents a second scenario: if Emma does very little revision before the exams, then a grade D is the most likely outcome. Once again there is some uncertainty, and she could score a grade C or E (based on what comes up on the paper on the day).

Try to write a comparison of the two trends shown in Figure 2.8 that mirrors the structure used in the story about Emma. Instead of talking about Emma's potential grade achievement, you will be writing about possible temperature increases (based on whether mitigation occurs or not). And instead of further uncertainty about what will be on Emma's exam paper, write about the further uncertainty surrounding *exactly* how much warming will occur for both scenarios (because the climate system is complex).

Thinking critically about geographical information

Figure 2.9 shows a choropleth map illustrating the global pattern of Gini coefficient values on a country-by-country basis. This is arguably a flawed resource.

- The Gini coefficient itself is a relatively complicated measure to understand (it shows the extent to which income is evenly distributed among the citizens of a territory: a Gini value closer to 1 than 0 indicates a higher degree of inequality).
- Because ten data classes have been used, the pattern is harder to analyse than it needs to be.
- In the study context of global disparities in economic development, it is unclear why anybody would want to produce a map with ten interval classes on it. Three or four classes might suffice if the objective (or 'story') of the map is to reveal broad global variations in the level of economic inequality found within countries.
- Can you suggest an alternative key for this map, using just three or four classes? What intervals might you use?

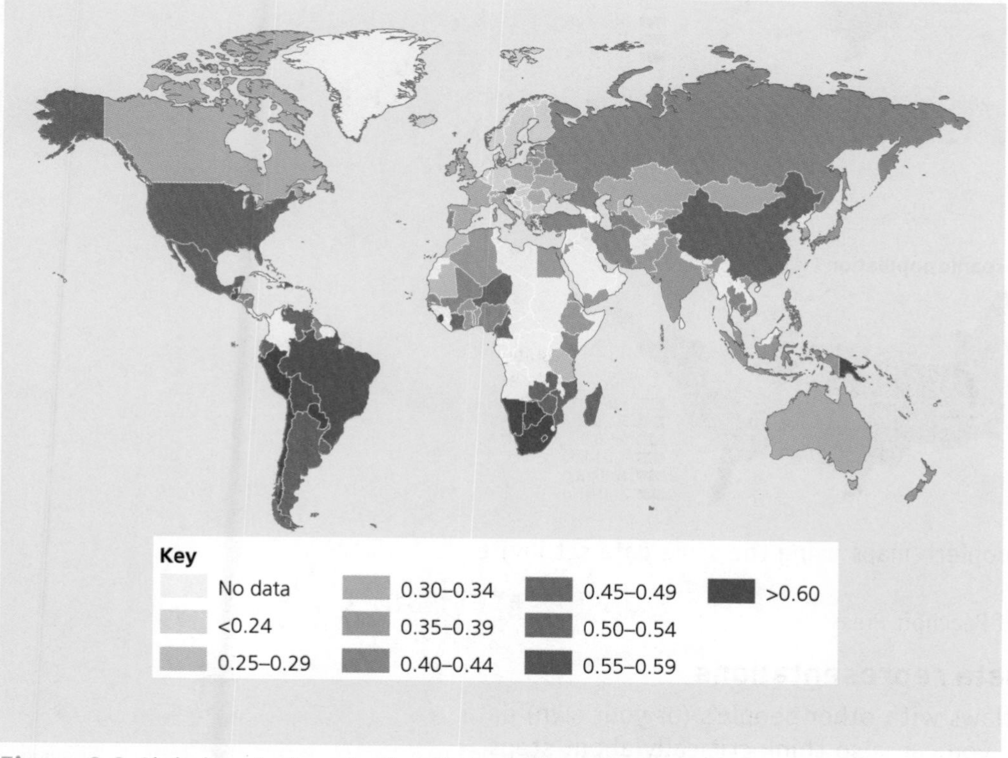

Figure 2.9 Global variations in the Gini coefficient, 2009

Take it further

Find other choropleth maps in your course textbook or using an online image search. Look for evidence of good practice and poor practice in the ways they have been created. In addition to the number of classes included, think too about how colour is used in different maps. Some unprofessional choropleth maps use too many wildly varying colours, which makes it harder to grasp immediately the pattern of 'highs and lows' (which are best shown with a gradation from darker to lighter shades of a single colour).

Figure 2.10 is interesting insofar as it shows two competing representations of the concentration of Hispanic people in different US states. As you can see, at face value the two maps tell different stories. One appears to show relatively low concentrations in most of the continental USA with the exception of one or two states. The other is telling us — at first glance — that much of the USA now has a very high proportion of Hispanic people living there. In fact both maps show exactly the same data. The only difference is that different class boundaries have been used to divide and map the data. This shows how easily facts can become exaggerated and in this context potentially 'politicised'.

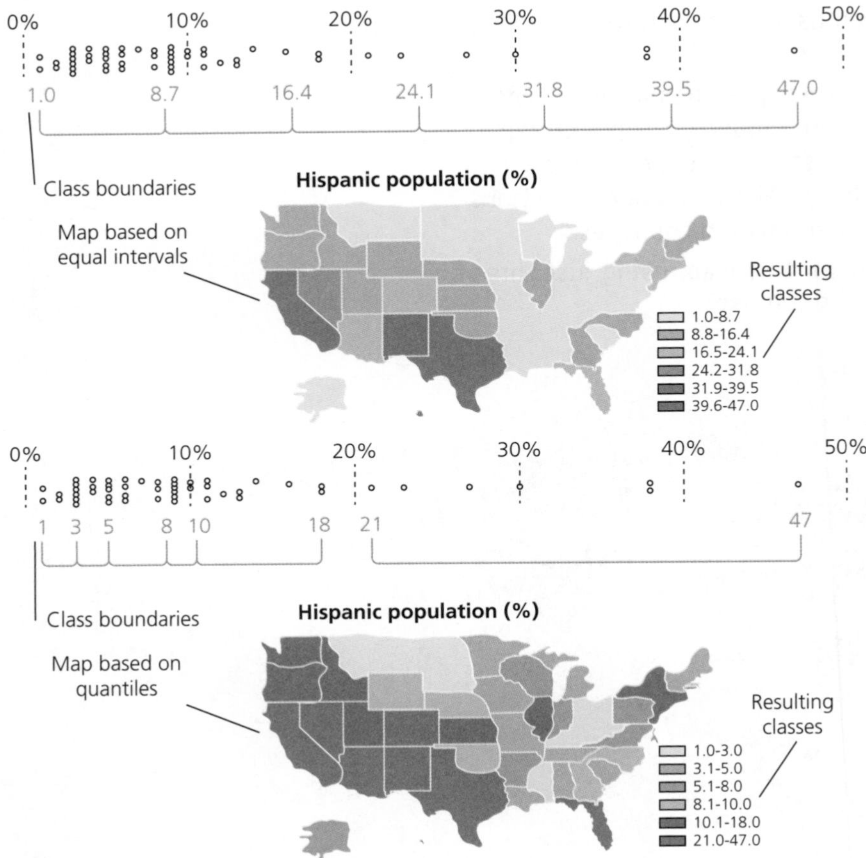

Figure 2.10 How two choropleth maps using the same data set invite differing interpretations

Reproduced by permission of Peachpit Press

Studying complex data representations

In addition to spotting flaws with other people's (or your own) data presentation techniques, you can also think critically about steps that could be taken to improve graphs and charts that, at first glance, appear perfectly sound, such as Figure 2.3 (page 28). In comparison, Figure 2.5 (page 29) offers a far richer representation of data because two patterns are shown and not just one. Not only are we shown how the number of migrants differs from place to place in Figure 2.5 (shown by the size of the circles), but we also learn how social attitudes vary spatially.

Even if complex data representations do not always feature in examination papers, you should make every effort during your 2-year course to improve your 'graphical literacy' by finding good examples to study. These may inspire you to produce more ambitious presentations of data in your own independent investigation (see Chapter 5). It is also good preparation for studying at university.

Annotated example

Bubble charts in geography are often associated with the late great professor Hans Rosling. His famous data presentations began with a simple scatter plot such as Figure 2.3 (page 28). Figure 2.11 shows how this simple chart (Figure 2.3) has been modified in several stages to create a richer data representation, which conveys multiple understandings about global inequalities in ways that are easy to grasp.

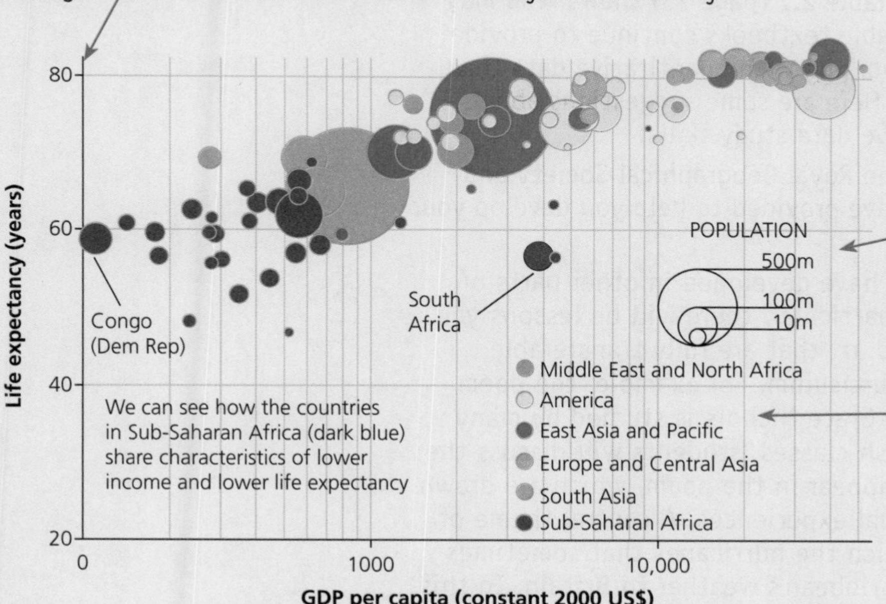

Figure 2.11 A modified scatter plot showing the correlation between GDP per capita and life expectancy
Reproduced by permission of the *Financial Times*

Now when we analyse Figure 2.11 we can say, for example: 'Many countries in Asia have similar high incomes and life expectancy to Europe, important exceptions being several with very large populations.'

In his lectures, Rosling also animated his graphs to show changes over time. You can view an example at: www.ted.com/talks/hans_rosling_shows_the_best_stats_you_ve_ever_seen

Take it further

The *Financial Times* and the *Economist* make frequent use of advanced graphical techniques. Be sure to look at these publications in a library or online, for example: www.ft.com/data-visualisation.

A good book to read about the use of data, charts and maps is *The Truthful Art*, by Alberto Cairo (2016).

Working with qualitative data

You are required to use and perhaps collect qualitative data while studying A-level Geography. This can include:

- interviews with members of the public or experts (which you may read about in articles and books; or may conduct yourself while carrying out fieldwork)
- novels, films, poems, music lyrics, diaries and other written texts (these may contain factual information but are first and foremost a vitally important data source for finding out about people's *perceptions* and *feelings* about places and environments)
- paintings and photographs (again, these can offer factual information about the way a place appeared in the past; but they also give us an idea of how the painter or photographer felt about the place they decided to memorialise)

The new courses, which began first teaching in 2016, put greater emphasis on the importance of qualitative data than A-level Geography did in the past, as Table 2.1 (page 23) shows. You may find, however, that some available textbooks continue to provide you with more assistance on working with quantitative data than they do with qualitative data. Here are some suggestions about ways to develop your qualitative data study skills:

- Use online resources that the Royal Geographical Society and Geographical Association have provided to help you develop your qualitative skills.
- Why not 'import' skills you have developed in other parts of the school curriculum? In particular, there will be lessons you learned in GCSE English and art that are fully transferable to the A-level Geography curriculum. For example, the poem 'Hurricane hits England' by Grace Nichols is studied by many UK GCSE students. In English classes, students will discuss the images and concepts that appear in the poem, which are drawn from the poet's own personal experiences. A central theme of this poem is the way in which the hurricanes that sometimes strike England bring the Caribbean's weather to Britain. In this way, two places become connected. At a more personal level, the hurricane led Nichols to reflect on her own West Indian heritage. There is considerable overlap here with the demands of the A-level Geography 'Changing places' topic, which requires you to explore cultural and artistic approaches to representing places.

- Make sure that you know how to carry out the coding of qualitative data. Here is an example to study: https://geographyfieldwork.com/CodingAnalysis.htm.

Activity

Review your own subject choices at GCSE and A-level. Where have you studied representations of places before? Were there any books or poems in your own GCSE English course that included place representations? How could you make use of this prior learning in your A-level Geography work?

Application to the exam

This chapter has explored a wide range of geographical study skills and techniques that you are expected to develop during your course. Is there anything else that can be done to maximise your chances of scoring high marks in your final examination when answering questions that ask you to manipulate or analyse data? Here are some final tips.

- Do not rush analytical tasks, even if they appear easy. Make sure you know *exactly* how many minutes are available per mark and allocate your time accordingly.
- Do not waste time repeating the question at the beginning of the answer space (see also page 48).
- Remember to include any units, where applicable, when completing a calculation (for instance, you may be calculating carbon flux *in grams of carbon per unit time*, or energy flows in *number of barrels of oil equivalent per unit time*).
- Resist the temptation to begin explaining trends and patterns if the question is asking you to 'describe' the resource.
- Do, however, take every opportunity to use appropriate geographical terminology as part of a well-informed description or analysis. You can see an illustration of how this can be done in the worked example overleaf.

Worked example

Below is a series of exam-style questions based on a resource (Figure 2.12) showing global flows of remittances.

(a) Calculate the range in values of remittance flows. **(2)**

(b) Analyse the relationship between the length and value of the remittance flows shown. **(6)**

(c) Explain *one* strength and *one* weakness of the data presentation methods used. **(4)**

(a) Lowest value is Japan to China: US$3 billion

Highest value is USA to Mexico: US$18 billion

Range is therefore 18 – 3 = US$15 billion

This must be measured accurately — don't rush the task!

Remember to include the units as part of the answer.

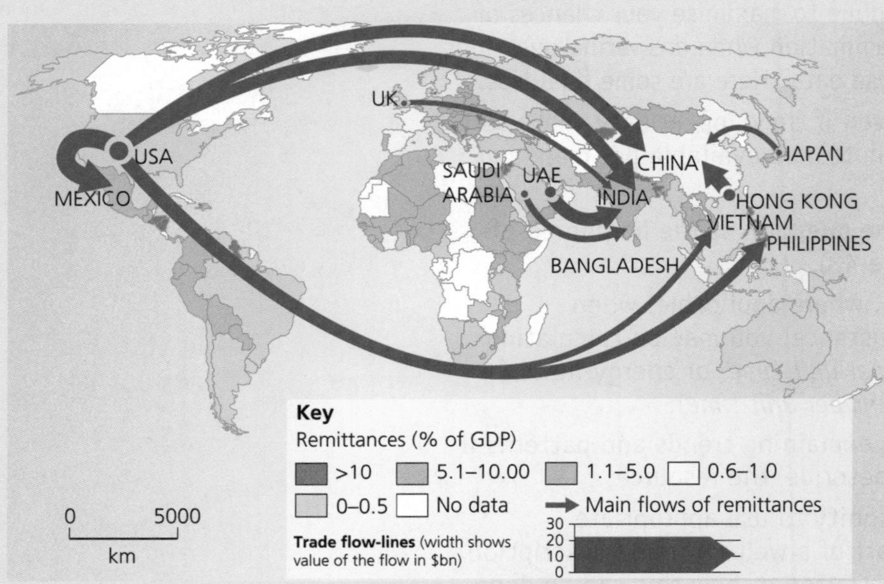

Figure 2.12 Flows of migrant remittances, 2011

(b) Overall, there is no particularly strong relationship. It is true that two of the very largest flows (from Hong Kong to China, and from the USA to Mexico) travel only short distances between bordering countries. However, the flow from Japan to China is also relatively short in length yet has a very low value. None of the four longest flows shown are quite as large as the flow from USA to Mexico. However, the flows from the USA to China and India are both around $10 billion in value — despite travelling enormous distances of around 30,000 kilometres. I therefore conclude that there is no particularly strong evidence of an inverse relationship between length and value. Instead, it appears that, in an interconnected world, large value flows travel very long distances as well as short distances.

It is good practice to offer an overview of what you are looking at in the very first sentence.

From the outset, this answer is focused on the apparent complexity of the relationship.

Examination mark schemes will sometimes award 1 mark for the use of data — you should always include a numerical value in an answer such as this.

It is very good to see the map's scale being referred to like this.

Although you are *not* required to *explain* the remittance flows, the inclusion of appropriate terminology here helps provide an analysis, which shows the student is 'thinking like a geographer'.

(c) Strength — this is a data-rich illustration because it combines a visualisation of remittance flows with the value of remittances for different countries (shown on the map). This is far more effective than producing two separate illustrations and also allows us to see the pattern of remittance flows arriving in countries with a relatively high dependence on them, such as Mexico and India.

Weakness — the use of six different classes of shading for the choropleth map makes the illustration more confusing to study at first than it needs to be. Three colours of shading — to show high, medium and low dependence — would be enough to get the message across effectively.

A proper explanation is being offered, which clarifies the points made in the first sentence.

If time allows, it is always useful to briefly include a further elaboration in order to maximise your chance of scoring full marks.

By offering an alternative and improved approach to the design of the map, the weakness is explained more fully.

You should know

- **Rushing perceived 'easy' mathematical and graphical tasks in your exams can result in marks being lost carelessly.**
- **It is not enough to be able to carry out statistical tests; sometimes the hard part is being able to communicate clearly what the results *mean*.**
- **Effective analysis of patterns, trends and data involves using the right words with clarity and precision. Getting it right can take practice.**
- **Apply your active reading skills to studying the illustrations in your textbooks. Thinking critically about the strengths, weaknesses and reliability of diagrams is good preparation for the exam papers and for writing your independent investigation.**
- **Make sure you are familiar with possible questions using qualitative data that could appear in your exam.**
- **To access the top marks in the exam and independent investigation, clear and coherent writing — including accurately spelt key words — is often essential.**

3 Knowledge-based, short-answer tasks

Learning objectives

> To understand how different kinds of knowledge-based, short-answer task are best answered
> To demonstrate succinct and effective communication skills
> To avoid common pitfalls associated with knowledge-based, short-answer tasks
> To learn strategies for success in the final examinations, under timed conditions

There are two kinds of knowledge-based, short-answer task included in your final examinations. These will also feature regularly in the written assignments your teachers will expect you to complete during the course of your studies. These are:

- A01-targeted short-answer tasks
- A02-targeted short-answer tasks

A01-targeted short-answer tasks require you to demonstrate knowledge and understanding. Examples of this type of question are:

Describe *two* hydrological flows. **(2)**

Explain how water is transferred from the atmosphere to the land. **(3)**

The second example, using the command phrase 'explain how', is asking for a slightly more detailed account than the first example. Essentially, however, all that is asked for in both cases is detailed recall of information.

A02-targeted short-answer tasks take things a step further by requiring you to apply knowledge and understanding. Examples include:

Explain *two* reasons why the amount of water stored in a drainage basin varies from season to season. **(4)**

Study Figure 3.1. Suggest why the percentage of rainfall that becomes overland flow varies during the course of the storm event. **(6)**

These kinds of A02 task require additional thought (some exam designers refer to these as 'multiple cognitive operations'). Not only do you need to recall information, but you must additionally apply it in a *selective* or *unexpected* way.

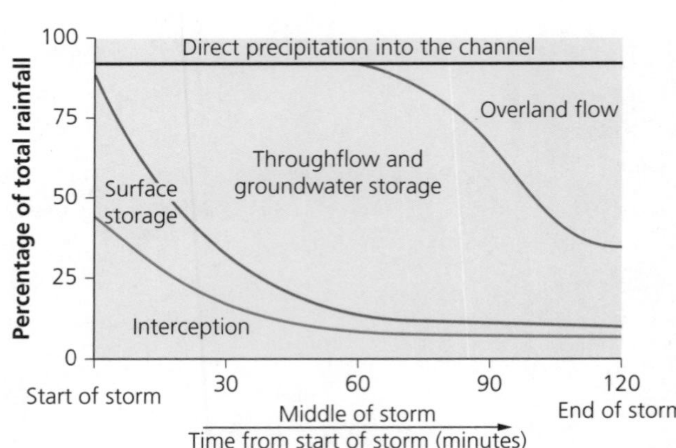

Figure 3.1 Changes in hydrological processes during a storm event

When answering the first AO2 question shown above, it is important to maintain a very tight focus in your answer on the key parameters of the question. Make frequent use of the words 'amount', 'storage' and 'season' in your response, even though it may be no more than 80–100 words long.

Alternatively, AO2 tasks can be linked to visual resources (such as the maps, graphs and charts shown in Chapter 2). The second example AO2 question shows this approach. When answering, you will need to make frequent references to Figure 3.1, perhaps quoting data. The approaches of the different examination boards do vary slightly in this respect so you will need to be familiar with the mark schemes used by your specification (see Chapter 6).

Table 3.1 shows a summary of the different kinds of short-answer task favoured by the different specifications. Approaches vary in terms of the command words typically used and also the number of marks and lines allocated in the exam answer booklet (for those exam boards that take this approach).

Table 3.1 Different approaches to short-answer tasks

Specification	Command word and main associated AO	Typical mark	Booklet answer lines
AQA	Outline (AO1) Explain (AO1)	4 4	8 8
Edexcel	Suggest (AO2) Explain (AO1 or AO2)	3–6 3–8	6–12 6–16
OCR	Suggest (AO2) Explain (AO2)	2–4 3–8	n/a n/a
WJEC/Eduqas	Outline (AO1) Suggest (AO2) Explain why (AO2)	3–8 3–8 3–8	n/a n/a n/a

Study skills

During your course, you can expect to complete many short-answer tasks as part of your homework or classwork assignments, end-of-unit tests and mock examinations. Your teachers may rely mainly on short-answer assignments during the first few months of your course (believing that it is better to wait a while before asking students to tackle more challenging extended essay writing). The advice given in this chapter is therefore likely to be relevant to the very first pieces of written work you complete and will hopefully help you to 'hit the ground running'.

Writing concisely

One of the most difficult things to gauge in the early stages of the course is the level of detail which is expected of you when completing a short-answer task. Consider, for instance, the question:

Explain *two* reasons why large numbers of people leave rural areas in developing countries. (4)

If 4 marks are available for this question, how long should the explanation of each reason you give be?

Most examinations use an answer booklet with lines provided for you to write your answer in, and this gives some guidance as to how long your answer should be. However, handwriting size can vary enormously among a group of students.

The best advice is to devote time to studying the mark schemes for past examinations, including any specimen papers. These provide you with valuable guidance on exactly what is expected. You do not want to do less than is recommended; it is also unwise to write a great deal more than is expected because this may jeopardise the amount of time remaining for you to complete other questions in the same examination.

When 2 marks are available for an answer, most mark schemes will award 1 mark for a 'basic' point and 1 further mark for a 'developed' or 'exemplified' point:

- A **developed** point takes the explanation a step further (perhaps providing additional detail on how a process operates).
- An **exemplified** point refers to a relatively detailed or real-world example in order to support the explanation with evidence.

Exam tip

When you are completing a short-answer task for homework, exercise the same judgement that you would in a real examination when it comes to the number of words you write. You need to master the art of writing succinctly while also scoring full marks. Producing a long answer instead of a short answer for the task your teacher has set you is *poor* preparation for the examination that awaits you at the end of the course.

The difference between...

Here are some possible responses to the following question:

Explain *one* reason why there is limited access to the internet in some countries. (2)

How to give a developed explanation	How to give an exemplified explanation
In some countries, people cannot use the internet because of political interference by the government. Citizens are basically not free to visit websites and if they do they could face severe punishment or imprisonment. *This answer simply states that some governments limit citizens' freedom. The second sentence is not a valid extension of the first point; it is merely 'common sense' writing with no real depth of knowledge or understanding. This only gains 1 of the available 2 marks.*	In some countries, average incomes are extremely low and most people living there cannot afford internet access due to poverty. A good example of this would be a very poor African country like Ghana. *This would be awarded the first mark for the basic idea of low incomes and poverty. In an A-level examination it is unlikely that the second mark would be awarded for simply naming an African country. This is not what is meant by exemplification. Only 1 mark is awarded so far.*
Some governments limit citizens' freedom to post information on the internet. In autocratic states, human rights like freedom of expression are not upheld the way they are in democracies. *This answer provides a valid extension of the first point using specialist terminology and demonstrating knowledge and understanding of the global governance topic. Many students would not be able to resist adding examples here, such as China and North Korea. However, enough has already been done to score both marks.*	Low incomes may explain the lack of internet access in the world's poorest countries. For example, GDP per capita remains below US$1000 per year in some of the poorest sub-Saharan African countries such as Somalia. *A second mark would be awarded here for sound exemplification. Can you see why? It is also important to realise that this represents sufficient exemplification; it would be a mistake to continue writing about Somalia (even if you know more facts). Enough has been done for full marks.*

Some short-answer tasks require you to develop an explanation in greater depth than the examples shown above. For instance:

Explain the concept of mass balance in relation to water stored in the cryosphere. (4)

This is an AO1 task because it does not ask you to 'explain *why*...' and therefore does not demand a *reasoned* response. Instead, all that is required is a series of statements conveying knowledge and understanding of mass balance and cryosphere storage. A mark scheme for this question would most likely take the form of a series of bullet points; a model answer might thus mirror these exactly:

- Earth's cryosphere includes ice sheets, valley glaciers and permafrost. ✓
- 'Mass balance' describes the way the amount of water stored in a system stays the same in the long run. ✓
- In the absence of climate change, we expect the amount of water stored in the cryosphere to stay the same from decade to decade. ✓
- However, there may be variations from year to year due to short-term influences such as sunspot activity. ✓

In contrast, look at the following AO2 short-answer task:

Explain why the growth of transnational corporations (TNCs) could lead to rising inequality within some countries. (5)

This question uses the command phrase 'explain why' and asks for an explanation that needs to be presented in *logical stages*. The level of demand is much greater than in the previous question because knowledge and understanding need to be *applied* carefully in order to answer a very specific and narrow question. A mark scheme for this question might be either points-based or levels-based (approaches vary for different specifications). Whichever approach is taken, maximum credit would be reserved for answers that:

- apply a range of accurate knowledge and understanding
- address the question directly, in a developed way

Figure 3.2 shows a sequence made up of three logical steps. Together, they provide the explanation that the question about TNCs asks for. At each step, knowledge and understanding are applied in order to make a point, and the point is also developed.

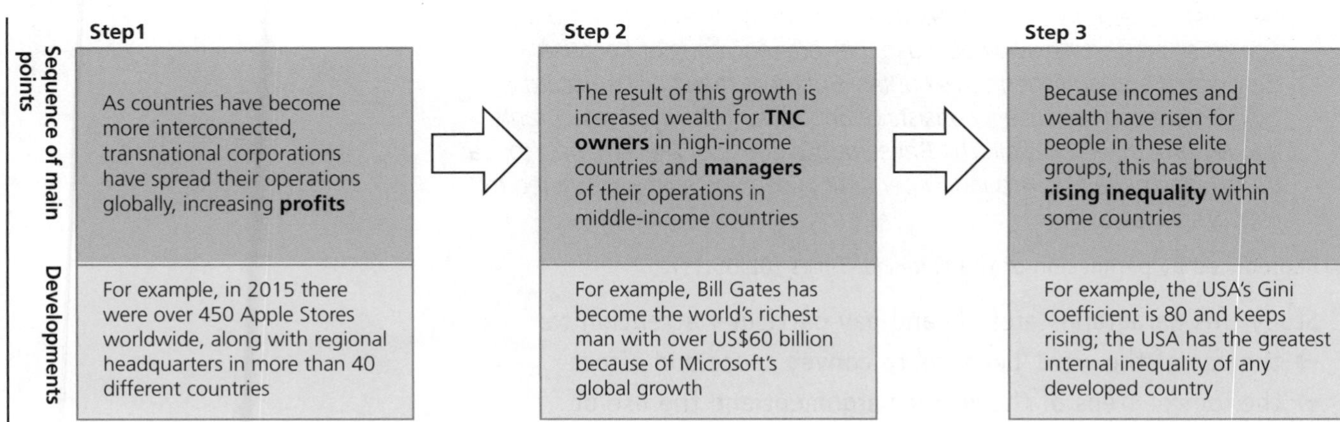

Figure 3.2 Structuring an answer

- If a points-based mark scheme were being used to mark this question, then three developed points have been provided (which would gain full marks because a developed point is worth 2 marks).
- If a levels-based mark scheme were being used, then the three logically sequenced stages that feature in this answer (each of which is developed and well-applied) should be sufficient for the top level.

Learning from the professionals

In addition to helping you gain knowledge, studying can help you become more proficient at writing concisely. Pay particular attention to the way professional writers use language and evidence. Look also at how they structure the points they make in a logically sequenced way. Think of it this way: if you play football or the guitar, it makes perfect sense to study the form of professionals whom you admire. You can identify and copy particular moves that they make. The same principle can be applied to writing — nobody is born good at it. Instead, we must all practise hard and develop our own style over time. When it comes to short-answer tasks, you can learn a lot from looking at how newspaper journalists in particular write. They have been trained to:

- explain what has been happening in a clear, concise, logical and well-ordered way
- use evidence to support each point made (in order to avoid any accusation that what they are saying is an opinion or assertion, and not factual)
- plan and write quickly in order to meet tight deadlines

Does this sound familiar? It should. It is exactly the same skill set that you are trying to develop.

The paragraph that follows originally appeared in an article in the *Financial Times* newspaper and is around 140 words long (around the same length that an answer to a 6-mark exam question should be). It has clear A-level curriculum relevance because it relates to global governance of 'the global commons'. The paragraph provides an explanation of why marine biodiversity is under threat in British waters.

> *Despite fishermen cutting the levels of discard in the industry by two-thirds during the past decade, there were still 51,179 tonnes of fish discarded by British fishermen last year, nearly a tenth of the total that was landed. Some fish are thrown away because they are too young or because they would exceed fishermen's quotas. But more than half of discard is typically wasted because it consists of unpopular fish which restaurants and fishermen will not sell. The British public does not like to depart too far from its favourite fish. Around 74 per cent of seafood sold is restricted to just 10 species.*

Study this paragraph carefully and pay particular attention to:

- the use of the word 'because' to convey cause and effect
- the logical steps of the writer's argument and the use of language to support it ('Despite...', 'Some...', 'But...')
- the sustained use of evidence to develop each point
- the clear subject-verb-object framework of each sentence
- the use of commas to help structure complex sentences

The difference between...

The table below provides a useful summary of techniques that tend to distinguish amateurish non-fictional writing from the more professional 'voice' of quality newspapers and academic textbooks.

Amateurish/immature tone	Professional/mature tone
Causality unclear	Good use of 'so' or 'because'
Narrative relies on simple techniques, e.g. an unstructured list of ideas	Logical sequence, with clear cause and effect
Disjointed arguments — the steps of each argument are not entirely clear or logically ordered	Sentences are logically connected, building an explanatory argument or discussion
Personal voice, asserting 'I think that...'	Impersonal use of third person ('The most important factor seems to be...')
Strong influence of the author's speech patterns on what is written, including use of slang	The style of writing is different from the writer's speech, and avoids slang and idioms
Assumes the reader has background knowledge of the subject	Assumes that the text is standalone and can be read by anyone as a 'voice of authority'
Repetition of sentence structure	Varied sentence structure
Repetition of words	May use synonyms
Unsure about capitals and full stops	Uses a range of punctuation correctly, for example colons
Unsure about use of apostrophes for possession and abbreviation, also about use of acronyms and dates	Correct use of apostrophes

Work presentation and handwriting

Many people would like to see all students being allowed to use a word processor in their examinations. Currently, however, only a minority of students can do so and it will most likely be many more years before universal 'e-assessment' becomes a reality. As a result, the quality of written work presentation and handwriting remains a concern for teachers, students and examiners alike.

This next section explores ways in which poor presentation skills can jeopardise attainment in short-answer tasks. Practical suggestions are offered for ways to address presentation shortcomings when completing written work as part of your ongoing studies (with an eye to perfecting your technique in time for the final examinations).

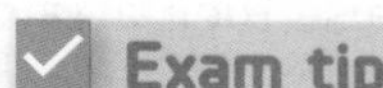

Exam tip

Think of your exam answer booklet as a 'conversation' with the examiner. In conversation, we are often influenced by other people's body language, and the way they dress (sometimes more than by what they say). Unfortunately, appearances always matter and some people do in fact judge a book by its cover.

Improving your presentation

Figure 3.3 shows an example of a handwritten short answer. It has been annotated to show how marks have been lost because of poor presentation.

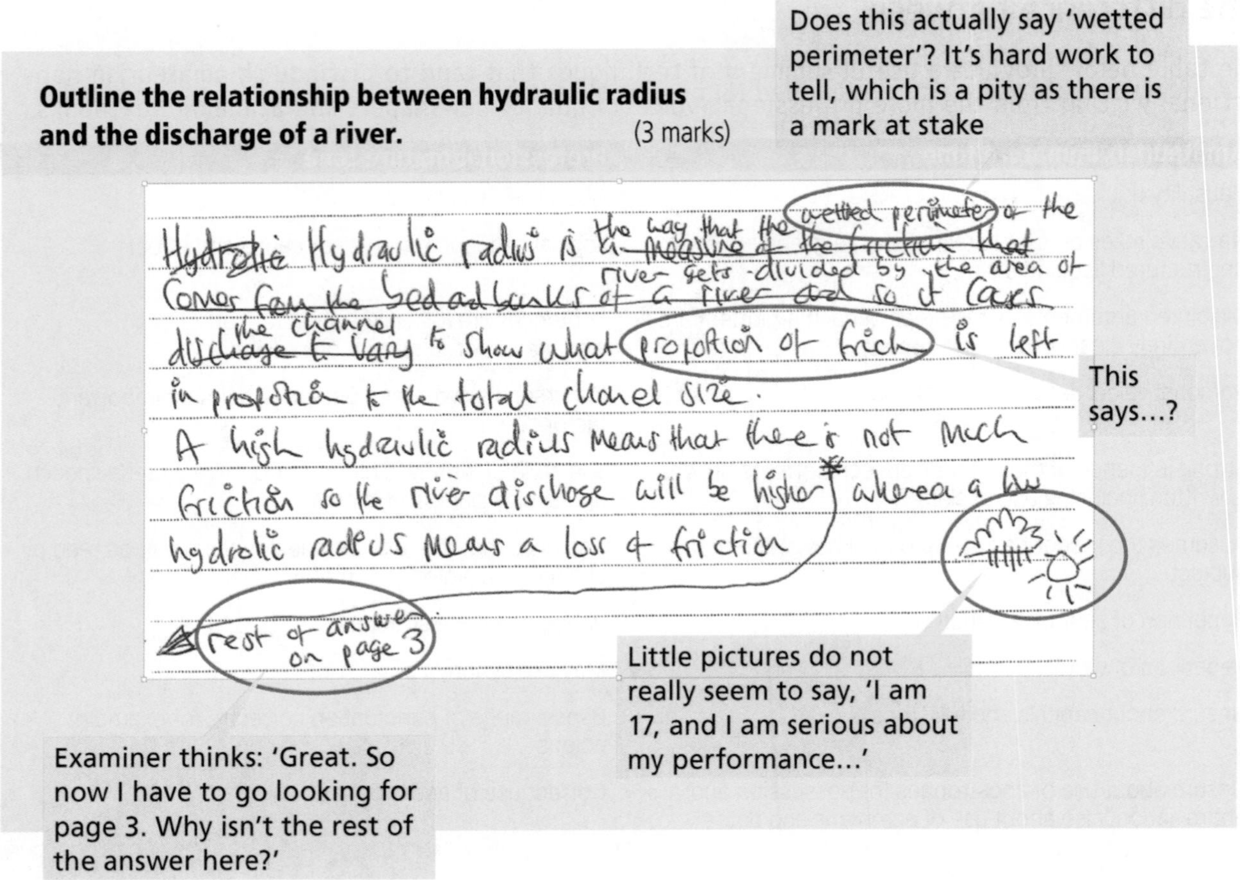

Outline the relationship between hydraulic radius and the discharge of a river. (3 marks)

Does this actually say 'wetted perimeter'? It's hard work to tell, which is a pity as there is a mark at stake

This says...?

Little pictures do not really seem to say, 'I am 17, and I am serious about my performance...'

Examiner thinks: 'Great. So now I have to go looking for page 3. Why isn't the rest of the answer here?'

Figure 3.3 Poor presentation

What easy, practical steps could this student take to improve their performance and raise their attainment? Advice we could give includes the following:

- Slow down a little when you are completing a written assignment. If writing under test conditions, think more and write less; try to get it right the first time you put it down. Avoid too much crossing-out and over-writing.
- Try to change your handwriting script. If teachers are forever telling you that your handwriting is too tiny to read, or is so large that bundles of extra pages always get tied to your exam scripts, then in either case, the teacher or examiner is having to work harder and longer to mark your work compared with that of other students.
- Practise writing with a pen as much as you can. You need to get used to writing for long periods of time, or else you risk finding it uncomfortable or even painful to use a pen. The result? Cramped, illegible handwriting. Do not type every homework essay if there are handwritten exams waiting for you at the end of the year. 'Use it or lose it,' as they say.
- When carrying out any activity, practice helps make perfect. Try to provide yourself with additional opportunities for writing with a pen whenever possible. This can include making handwritten notes when reading articles and carrying out research (see Chapter 1) or keeping an old-fashioned appointment diary

(instead of relying on your phone for everything). In many ways, it goes against the grain to be rejecting modern technology in favour of ink and paper. But the reality is that you and other students will continue to be examined in a traditional way for the foreseeable future. You would therefore be wise to adopt some traditional training methods.

Signposting key information

If your writing is very, very hard to read, then you run the risk of not getting the reward you deserve for your written work because your teacher or examiner has not given you sufficient credit. Do not let important terminology vanish into a vortex of cramped writing, crossings-out and ink smudges. Take care to print very important key words — that you strongly suspect will trigger marks — in CAPITALS, or perhaps underline them instead. This is called 'signposting' key words and maximises your chances of gaining the credit you deserve. Figure 3.4 shows an example of handwriting that is generally untidy, but where due care has been taken to carefully print and signpost key words.

> Another important influence on globalisation are TRANSNATIONAL CORPORATIONS (TNCs). These are massive companies whose HQs are usually in MEDCs or EMERGING ECONOMIES. TNCs gain access to different markets using a range of strategies such as GLOCALISATION (which means changing the product to take into account local CULTURE).

Figure 3.4 An example of generally untidy handwriting but with carefully printed key words

Some suggestions for improvement

Finally, it is a good idea to continually reflect on your performance and set yourself improvement targets if necessary. Hopefully, your teacher will assist you with this, perhaps by identifying one strength and one weakness in every assignment you complete. Another approach that you can adopt yourself is to briefly make notes about each piece of written work your teacher marks for you as follows:

Title: ______________________________

Mark: _____

One thing that I did well: ______________________________

Two things that I can improve:

1 ______________________________

2 ______________________________

For this to be an effective self-appraisal, you cannot just write things like 'I got a high mark' or 'I need to write more'. Instead, the comments should be more along the lines of 'maintained my focus throughout' and 'need to include more evidence or place names'.

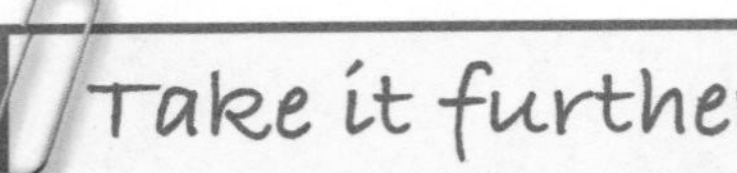

Take it further

Spend some time looking through Manchester University's 'phrasebank' for some ideas on connectives and sentence stems that you can integrate into your work: www.phrasebank.manchester.ac.uk.

If you score particularly low marks on a written assessment, you may want to consider resubmitting it after you have reflected on where you went wrong. Most teachers are impressed when students do this because it shows that you are really committed to trying to improve your performance.

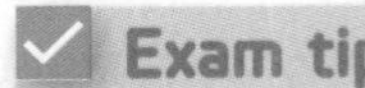

Exam tip

Read your answers back to yourself before you hand them in. It is remarkable how many grammatical errors or theoretical misunderstandings you might detect when you do this. This proofreading tip applies to people at all stages of their career.

Application to the exam

This chapter has provided you with a range of suggestions for ways to improve your technique when tackling short-answer tasks. When the time comes to apply all you have learned to the exam there is, of course, an additional challenge to contend with — *writing under strictly timed conditions*.

Writing under timed conditions

It is a fact of life that some people can write very quickly. If you are one of those at the other end of the spectrum, do not panic; there are practical steps you can take to adapt:

- First, it is essential to acknowledge this particular constraint on your performance and to adopt a strategy that will help you adapt to the challenge (see below).
- Make sure that you know *exactly* how many minutes you have to complete each question. You must be extremely disciplined and time yourself strictly. Move on to the following question as soon as your time is up.
- The next step is to quickly form an overview of each question you encounter and identify what the main ingredients of success are going to be (try to anticipate what the mark scheme will look like). For example, you need to form a rough idea of how many logical and developed steps need to be included when tackling an AO2 short-answer task, such as the one shown on page 43.
- Make sure your answer includes each logical step that needs to be there, even if you need to resort to using bullet points, abbreviations and 'streamlined' grammar.

Exam tip

Whatever you do, don't waste time repeating the question at the start of your answer. Try also not to repeat yourself by basically saying the same thing twice but in slightly different ways.

Using abbreviations, acronyms and shorthand symbols

Table 3.2 shows commonly used symbols, acronyms and abbreviations that most geography teachers and examiners will be familiar with. In short-answer tasks, no credit is usually reserved for spelling, grammar or the quality of written communication. As a result, the use of abbreviations, acronyms and symbols such as those shown here is unlikely to result in your work being penalised.

Table 3.2 Commonly used symbols, acronyms and abbreviations

Symbols		Acronyms		Abbreviations	
=	Equals	GDP	Gross domestic product	Govt	Government
>	Greater than	HDI	Human development index	Sust devt	Sustainable development
<	Less than	HIC	High-income country	Popn	Population
e.g.	For example	NGO	Non-governmental organisation	Mass mvt	Mass movement
→	Leads to	FDI	Foreign direct investment	Chem	Chemical
∴	Therefore	TRF	Tropical rainforest	Atmos	Atmosphere
∵	Because	NPP	Net primary production	Hydro-met	Hydro-meteorological

Activity

What other abbreviations, acronyms and shorthand symbols could you use when under pressure in an exam? Go through each topic you have studied and list suggestions.

The difference between...

The two examples of student answers shown below are, in essence, identical answers to the question:

Explain the economic and social consequences of high levels of in-migration for urban areas. **(4)**

Longhand example	Shorthand example
There are many economic and social consequences of migration from urban areas. Economic consequences can include a greatly increased workforce, which can increase the profitability of urban businesses, leading to economic growth, measured by an urban area's contribution to the country's gross domestic product. For example, London's economy has grown very strong since eastern European migrants began arriving there after 2004 when the European Union was enlarged. The social consequences can be positive, such as increased cultural diversity. London now has many Polish delicatessens, which people from other ethnic groups enjoy using too. There are negative social consequences as well, though. Very high levels of in-migration can put pressure on housing, and inner London rents have skyrocketed in recent years. This can result in a lack of affordability and the growth of homelessness.	Economic — greatly increased workforce → high *profitability* for businesses and ∴ an urban area's contribution to *national GDP* may increase (e.g. London, since EU enlargement and arrival of Polish workers in 2004+). Social — positives include greater *diversity* (e.g. adoption of Polish food in other ethnic neighbourhoods). Negatives include rising rents because of population pressure → lack of affordability and *homelessness*.

If you are worried about completing all of the questions in the exam under timed conditions, can you see the advantage of adopting a style similar to the second example in some of your answers?

You should know

- 'Maxing out' your marks on short-answer tasks is a good strategy to increase your final grade, so make sure you have good exam technique.
- Make sure that you have a good feel for how much needs to be written to gain 'extension' or 'development' marks. Don't write too little, but it is equally important not to write too much and run out of time for answering other questions.
- Take note of how good journalists write and try to adopt the same 'punchy' style when tackling short-answer tasks in your exam.
- Remember that appearances matter — someone with A-grade ideas but whose work is almost impossible to read can end up with a C-grade.
- If you cannot write very quickly, adopt strategies (abbreviations and symbols) that will help you convey what you really need to say in the examination using as few words as possible.

4 How to write an evaluative essay

Learning objectives

> To understand how essay-writing skills can be developed during the 2 years of full A-level study
> To learn the difference between descriptive and evaluative essay writing
> To explore possible approaches to writing a conclusion and making a final judgement in an essay
> To discover how to succeed in essay writing in your final examinations under timed conditions

This chapter is devoted to the study and examination skills needed to support **evaluative essay writing.**

- Evaluative essays can be a demanding form of written assessment and are an important driver for the depth of teaching and learning required for different topics.
- Some students worry more than they need to about the perceived difficulty of evaluative essay writing. There are several easy-to-adopt techniques that can provide you with the 'scaffolding' you need to carry out these extended-writing tasks confidently.

Different specifications favour particular command words or phrases for essay tasks, such as 'discuss', 'assess', 'evaluate' or 'to what extent'. Make sure that you are familiar with the terms used most commonly as part of your own specification's examined assessments. The precise number of marks available (or 'tariff') for essays varies according to the specification you are studying, as does the exact time allocated to the task. As a broad rule, however, the following generalisations apply:

- A-level Geography students are required typically to write evaluative essays that, under timed conditions, must be completed in around 20–25 minutes.
- In general, this represents around two or three pages of average-sized handwriting, i.e. somewhere between 600 and 800 words.
- In order to score highly, evaluative essays should incorporate both an introduction and a formal conclusion.

Some A-level Geography specifications include an even longer evaluative essay-writing task of 30–45 minutes (see Chapter 6).

Activity

As either an individual or class activity, compare the essay assignments for your geography course with those of your other A-level subjects (where applicable). How similar are the essay-writing requirements for history, economics, English or politics, for instance?

At the outset, there are two important observations to make about evaluative essay writing at A-level. Firstly, the level of demand is considerably greater than for extended writing at GCSE.

- Do not worry if at first you find that essay researching, planning and writing is a tough challenge. A-level Geography, in common with other subjects, has been designed as a 2-year programme of study during which time all students mature as learners.
- Essay-writing skills do not blossom magically overnight. Like any other skill — from driving a car to learning to play a musical instrument — essay-writing prowess builds gradually over time through trial and error.
- Most teachers will grade the first essays written by their students at the start of Year 12 using real A-level mark schemes. Remember that those mark schemes have been designed to assess the geographic capabilities of 18-year-old students who have completed their Year 13 learning. It is therefore unsurprising if many Year 12 students score mediocre marks for the first few essays they complete as homework or classroom tests.

Secondly, at the start of their course, many A-level students mistakenly think that an extended-writing question — especially one with a high tariff and a large space to fill in the answer booklet — is an open invitation to 'write all I know' about a particular topic. It is not, however, as this chapter demonstrates.

- In Year 12 it is still common to see students submitting word-processed homework essays for grading by their teachers that contain way too many lengthy and overly descriptive case studies. Later on in the course, most Year 13 students will start to think more carefully about the quality — rather than quantity — of the words in the essays they submit for grading. This is due mainly to their growing familiarity with essay mark schemes. In particular, students gain an awareness of the need to meet the different assessment objectives that feature in essay mark schemes.
- Most important of all is the distinction between AO1 and AO2 content, which is used when essays are being graded. The former consists of descriptive information (including definitions of technical terms and recalled information about places, processes or case studies). The latter includes evaluative writing, which is characterised by a discursive, argumentative or critical tone, and which may show awareness of contrasting perspectives in a particular geographic issue, theory, idea or place. Failing to grasp the AO2 requirements of the assessment can result in an able student gaining no more than half marks for a well-researched, yet overly descriptive, essay.

The difference between...

Descriptive or explanatory extended writing	Evaluative essay writing
This usually consists of factual points that: • outline the main features of a place, object, issue, idea, theory, strategy, etc. • recall why these features have arisen because of particular factors, processes, actions, management objectives, etc. For example, the explanatory statement: *There are now more than 7 billion mobile phones — more than one per person. This number has risen rapidly in recent years. Just 10 years ago, there were fewer than 1 billion mobile phones. Mobile phones and other technologies are an important factor contributing to globalisation. By having a mobile phone, people can experience the 'shrinking world' and so become more globalised.*	This consists of material that may additionally: • reflect on the limitations of a particular explanation or perspective on an issue • discuss spatial or temporal variations in the operation of factors and processes • use important geographic concepts, such as scale, or establish connections and relationships between different ideas For example, the evaluative statement: *There are more than 7 billion mobile phones on Earth — more than one per person. By having a mobile phone, people may experience the 'shrinking world' and become more globalised. However, many mobile users do not have access to the internet and just make local calls. In some places, mobile phone growth may not necessarily be linked with globalisation.*

Activity

Review the assessment objectives for each of your different A-level subjects (these will be included in the specifications, usually in a table after the subject content). What similarities and differences can you observe between geography and the subjects you are taking that are closest in character, such as history, economics or biology?

Study skills

During the 2-year geography course, plenty of your study time is likely to be devoted to researching and writing evaluative essays. Steadily, you will gain an understanding of:

- the necessary depth in which examples and case studies should be investigated
- the ways in which the information you collect needs to be adapted in order to meet the requirements of an evaluative essay task

As a general rule, you will be producing essays that are structured as follows:

- **Introduction** In a 600–800-word essay, the introduction should probably not exceed 100 words and is more likely to be around 50 words in length. Three or four sentences can be sufficient to provide a concise definition of any key terms and to establish a **focus** and an **enquiry framework.**
- **Content** A series of roughly equal-sized paragraphs should guide the reader through a sufficient number of themes or examples for a credible conclusion or final judgement to be made. For example, a physical geography essay may ask you to 'discuss the importance of erosion in the development of landforms over time'. A sensible approach might be to write four or five 100-word paragraphs, each of which deals with a different landform (on the basis that four or five landforms is a sufficient number for credible generalisations to be made).

→ **Conclusion** This should address the question directly and make a final judgement about the ideas and issues which the essay has focused on. It is advisable to produce a substantial conclusion — and not merely a perfunctory sentence or two. This is because mark schemes for evaluative essays usually put a high premium on the ability of candidates to (i) think critically and (ii) make connections between different geographic ideas and issues in order to develop a topic overview (see page 9). The clearest way to demonstrate to your teacher or an external examiner that you can meet these requirements is to write a conclusion which shows you doing exactly so!

The **enquiry framework** is an outline of the 'building blocks' for an evaluative essay. A brief summary of the enquiry framework can be included in the introductory paragraph. For instance, consider the essay question:

Discuss the view that barriers to globalisation are increasing over time.

Ideally, the building blocks for this essay are:

→ arguments both in support of and in opposition to the statement
→ evidence drawn from a range of spatial scales (given the fundamental importance of scale as a concept in geography)

An effective 65-word introduction to this essay might therefore be:

Globalisation is the increasing interconnectedness of people and places. It is a process with many economic, social, cultural and political dimensions. This essay will explore examples of increasing barriers to global flows, but also contrasting examples of falling barriers. Barriers to globalisation are introduced or removed not only by governments but by citizens and organisations too, and this essay will discuss examples at varying scales.

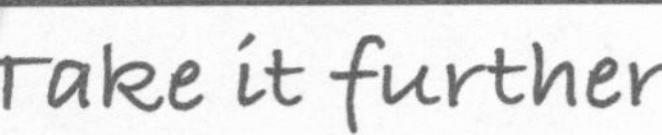

Pay attention to the way in which short columns and 'opinion pieces' in quality newspapers and magazines such as the *Financial Times*, *Guardian* or *Economist* are written. Find a recently written short column on a topic that interests you (no more than 1000 words in length). Count the number of paragraphs and estimate the number of words per paragraph. Look critically at the balance of detail (evidence and data) and argument that each paragraph contains. Could this serve as a template for your own evaluative essays?

Subject knowledge and evaluative enquiry

A-level Geography subject content differs from GCSE geography content not just in terms of the level of detail that is included, but also in the way it is written. A-level specifications tend to be phrased in ways that invite ongoing critical reflection and evaluation. For example, consider the four examples of A-level subject content dealing with global systems and globalisation shown in Table 4.1. In each case a clear signal has been given that teaching and learning about this topic *should progress beyond*

! Common pitfall

Throughout your 2 years of study, it is advisable to write essays for homework that are roughly equal in length to those that you are able to produce under timed conditions. Producing a homework essay that is several thousand words long — and very richly detailed — does not help you perfect the succinct style of evaluative writing needed to perform well under strictly timed conditions. Think of it this way: an aspiring athlete is unlikely to win many 400-metre races if they train over 5-km distances. A writer who has been employed to write 'punchy' 1000-word news items for a website needs to adopt a very different style from someone who writes 5000-word magazine articles. It takes a particular skill to produce a punchy and succinct 1000-word column. Practice is needed.

being able to describe and explain isolated ideas and examples. In particular, you are often guided by your specification to think critically about:

- the *importance* of the different issues, ideas and places that you study
- the possible *connections* between different topics and sub-topics

Table 4.1 Specification content dealing with global systems and globalisation

Exam board	Example of specification content	Content focus	Specification steer towards critical thinking
Edexcel	'Special economic zones, government subsidies and attitudes to FDI have contributed to the spread of globalisation into new global regions'	Globalisation and special economic zones	The specification wording implies that you need to do more than just learn how special economic zones operate. In addition, you should evaluate the importance of the overall *contribution* that special economic zones have made to the spread of globalisation (perhaps in comparison with other contributing influences, such as technology).
OCR	'International trade has increased connectivity due to changes in the 21st century, including the role of regional trading blocs, such as the EU'	Trade and trading blocs	The specification wording indicates that you might evaluate the *role* (or importance) of trading blocs as an influence on international trade (and not merely learn how to describe the operation of trading blocs using examples). The study of trading blocs should therefore incorporate critical thinking about their overall role or importance.
AQA	'The nature and role of transnational corporations (TNCs), including their spatial organisation, production, linkages, trading and marketing patterns'	Transnational corporations (TNCs)	Again, the specification wording prompts you to evaluate the importance of the role that TNCs have played in the growth of global systems (in comparison or in conjunction with trading blocs, for example). Here, the specification indicates that it is not enough to simply learn facts about the sizes of operation for selected TNCs.
WJEC/ Eduqas	'Flows of money, ideas and technology, linked with economic migration, that reduce or exacerbate global economic inequalities.'	Economic migration and global flows	The specification wording lends itself to a judgement of the extent to which migration reduces or exacerbates global inequality.

The table shows how content has been phrased deliberately in ways that actively stimulate evaluative enquiry. You are required to do more than learn about examples of trade blocs or TNCs. You are expected to study the contribution, role or importance of different factors in the growth of globalisation and the reproduction of global systems. This sends a strong signal that, as you research examples of trade blocs and TNCs, you need to be thinking critically about their importance.

Activity

Use the information in Table 4.1 to develop your own evaluative essay titles for the content shown. Along with a clear topic focus (such as TNCs), make sure that your titles use words such as 'role', 'importance' and 'contribution'.

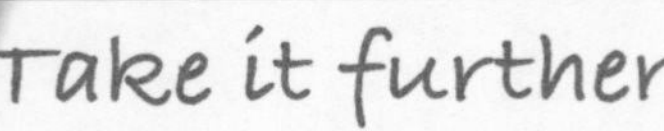

Review other A-level topics in your specification. Look for evaluative words such as 'role' and 'influence'. Can this provide you with possible clues for future essay titles that might appear in an examination?

Table 4.2 offers advice on how to set about studying two further geography topics. In both cases, when researching these themes it will be important for you to distinguish between *factual* information (data about place regeneration and water cycle flows, respectively) and *evaluative* material (points dealing with their success and importance, respectively).

Table 4.2 The study skills needed for successful evaluative essay writing

STUDY TOPIC		STUDY SKILLS
The success of place regeneration schemes	**The importance of different water cycle flows**	
You might research: • located examples of regeneration schemes • data showing the costs and investments made • data showing visitor numbers, or the number of jobs created by the schemes • other place impacts, such as environmental harm or improvements • viewpoints articulated by different people or organisations	You might research: • the flows that feature in different water cycle models • ways in which different flows operate and transmit water • rainfall or river flood data in different geographical contexts • ways in which water cycle flows impact on human life	Researching factual knowledge and case study details (AO1 — knowledge and understanding)
You might look for evidence of whether: • the benefits outweigh the investment costs or not • perspectives differ on whether a scheme was successful (and reasons for any differing perspectives) • economic success has brought social or environmental costs (thereby making the overall level of success hard to judge) • enough time has passed yet for a scheme to be judged as successful (and sustainable)	You might look for evidence of whether: • some flows are more important than others in particular contexts (because of climate or other factors) • certain flows, such as groundwater movements, are especially important for particular user groups, such as farmers • water cycle flows could start to become more or less important over time (perhaps because of cyclic or long-term climate change, or human intervention)	Thinking critically about information you have collected (AO2 — analysis and evaluation)

How much factual knowledge should an essay include?

When working on an evaluative essay assignment, try to judge sensibly how much detailed information you should be collecting for any single case study, place, landform, issue or theory. There is a limit to how much AO1 credit can be gained from the inclusion of excessive amounts of information and data.

As Chapter 1 suggested, it is important not to get too carried away making detailed notes drawn from lengthy online articles that only support a minor point in your geography specification (and by extension in an essay that you are asked to write). Try to be disciplined when taking notes: perhaps a one-paragraph summary of the article will suffice.

We can reflect on how much detailed knowledge a student would need to acquire from reading and research in order to tackle the following evaluative essay questions (assuming that the target length is, in each case, 1000 words or less):

- **To what extent are processes of deposition beneficial for human activity in different coastal environments?**
- **Assess the economic impact of international migration for different countries.**

In both cases, a good answer would ideally consider a *range* (three or more) of different coastal environments/countries affected by deposition/migration (rather than just one or two examples). In reality, this might mean that a maximum of 200–300 words would need to be written about any single coastal case study or country. Realistically, it is not necessary to research and remember very long lists of case study information. Instead, a handful of key facts (perhaps just three or four) is required, each of which offers an opportunity for evaluation.

Being able to judge *what to include and what to leave out* when researching and writing about a case study is an important skill that students develop over time. In general, the most important pitfalls to avoid are over-loading your essay with too much of the following:

- Background information: for instance, you may decide to use the USA as an example in the migration essay shown above. It is not necessary to begin your paragraph dealing with the USA by offering a historical overview of the discovery of America by Christopher Columbus. Interesting, perhaps, but this material will do nothing to help you develop an argument (which is the whole point of writing the essay).
- Irrelevant information: the focus of the migration essay is economic impacts. There is no point whatsoever in adding superfluous information about cultural changes resulting from international migration (unless you work very hard at showing and stressing that there are knock-on economic impacts).
- Excessively detailed information: in the coastal example shown above, it is worth reflecting on how much scientific detail needs to be included about, say, rates of sedimentation or fine particle flocculation in coastal estuaries. This is a geography essay, not a physics essay. One or two detailed points incorporating some data and technical terminology is usually sufficient to establish your authority when writing about a topic. Beware of including so much detail about physical processes (drawn from whichever academic articles or textbooks you have consulted) that you have no room left to actually answer the question — which is asking you to form a view about whether these processes are *beneficial* for humans or not *in different geographical contexts*.

! Common pitfall

Including material that does not help develop your answer wastes time and may mean that you fail to get across all the points that you need to in the space and time available for you to complete the task. A good rule to follow is to ask yourself whether each fact you include is 'need-to-know' information or 'nice-to-know' information. If it falls into the second category then leave it out! The critical point here is to manipulate your material in the exam so as to answer the question directly. This is when the decision on what to include and what to leave out is really significant.

→

Annotated example

What does a 'proper' evaluation look like?

Chapter 1 looked at active reading strategies and ways of 'deconstructing' a text passage in order to separate out its descriptive, explanatory and evaluative elements. Mastering this technique is important when using a range of case study texts in preparation for completing an evaluative essay assignment. For example, imagine that you have been asked to complete for homework the following essay:

Evaluate the economic impact of new flows of money and investment on different places you have studied.

The text below is taken from a newspaper article documenting the impact of new investment by the BBC in Salford (Greater Manchester) since 2004. The text has been annotated to show:

- descriptive information that can be included more or less verbatim in this essay (in order to gain A01 credit)
- information that can be thoughtfully adapted to provide an evaluation of the economic impact of investment by the BBC (in order to gain A02 credit)

Impact of BBC investment on Salford may be less than thought

The BBC moved a large part of its operations from London to Salford (Greater Manchester) in 2004. By so doing, the BBC hoped to reduce the costs and improve the quality of its shows. At the time, it was widely anticipated that this relocation would provide major economic benefits to Salford. However, a new report by Centre for Cities says that economic benefits have in fact been difficult to identify, and the effect of the BBC move on local employment has been negligible. Predictions that the BBC relocation would create 15,000 new jobs have not been met, according to Centre for Cities. In reality, only 4400 new jobs were created between 2004 and 2016. But Salford City Council (SCC) disputes the findings of the Centre for Cities report. An SCC spokesperson said the BBC's move north had sparked a 'creative and digital revolution in the region' worth £3.1 billion to Greater Manchester's economy. SCC believes that 55,000 jobs have been created and that the new report is flawed because it does not properly calculate the wider economic benefits of BBC employment.

A description of the investment made by the BBC (A01)

Evidence that the impact of the investment has been less than expected (A02)

Factual evidence of the impact of investment on employment in Salford (A01)

Evidence that the impact of the investment is disputed — which complicates the evaluation (A02)

Recognition that the complexity of the economic changes brought about by investment and money flows can make it difficult to weigh up the overall impact (A02)

Making a judgement when concluding

Evaluative essays may require that a judgement is made for part of the AO2 assessment. For instance, in both of the following titles you are asked to make a final judgement about the extent to which the viewpoint given is true or not.

- **'Vulnerability to tectonic hazards has risen over time.' Discuss this viewpoint.**
- **To what extent do you agree with the view that global governance of Antarctica has been successful?**

When writing a conclusion for either of these titles, it is vital that a final judgement is provided. Do not just sit on the fence. Draw on all the arguments and facts you have already presented in the main body of the essay, weigh up the entirety of your evidence and say whether — on balance — you agree or disagree with the question you were asked. To guide you, here are some simple rules that it is worth adhering to:

- Never sit on the fence completely. The essay titles you will encounter in geography examination papers have been created purposely to generate a discussion that invites a final judgement following deliberation. Do not expect to receive a high mark if you end your essay with a phrase such as: 'So, all in all, the governance of Antarctica has been both successful and unsuccessful.'
- Equally, it is best to avoid extreme agreement or disagreement. In particular, you should not begin your essay by dismissing one viewpoint entirely, for example by writing: 'In my view, the global governance of Antarctica has been entirely unsuccessful and this essay will explain all of the reasons why.' It is essential that you consider different points of view. The answers that score highest are likely to be well balanced insofar as roughly half of the main body of the essay will consider ideas and arguments supporting the statement, and the remaining half will deal with counter-arguments.
- An 'agree, but...' or 'disagree, but...' judgement is usually the best position to take. This is a mature viewpoint, which demonstrates that you are able to take a stand on an issue while remaining mindful of other views and perspectives.

Extreme agreement	Agree, but...	Sitting on the fence	Disagree, but...	Extreme disagreement

Figure 4.1 The spectrum of judgements

Source: Cameron Dunn

Take it further

Review some of the levels-based mark schemes for essays in the geography specification you are following. You may see words like 'balanced' or 'unbalanced' featuring in some of the levels. This indicates that you are expected to give *equal consideration* to different viewpoints about the statement you have been asked to write about (and so must not entirely dismiss one viewpoint out of hand).

Exam tip

Mark schemes for evaluative essays in human geography will often state that AO2 credit can be given to answers that acknowledge the existence of **contrasting perspectives** on the issue under discussion. This is particularly true of the 'Changing places' core topic (across all exam boards). Management decisions affecting cities or rural settlements are often controversial; it may be difficult to evaluate the overall balance of winners and losers, especially as time passes. In particular, gentrification is a process of change that divides public opinion. Many people may benefit from the gentrification of a place, particularly home owners and new property developers. However, poorer residents who rent their homes are often displaced from a place when gentrification occurs. They are far less likely to view its effects positively. The OCR exam board in particular puts great emphasis on students' ability to view issues from varying perspectives

There is a very good reason why A-level Geography students are required to make a judgement as part of their assessment. Being able to justify a particular course of action can be a vital workplace skill. Looking forward into the future, you may one day find yourself working in an environment where you have been given responsibility over a budget, or are managing other people. You might need to hire or fire workers, or decide which projects your company should invest money in or best avoid. Perhaps you will have to rule on whether a business client or customer is entitled to a refund or compensation. Whatever decisions you are required to make, you could also be asked to justify them to whoever, in turn, manages you. It is therefore very important to be able to explain exactly why you arrived at a particular decision. Which evidence or argument swayed you most, and why? Real life is full of difficult decisions that need to be reflected on and thought about carefully before a course of action is chosen.

Exam tip

Try to include *ongoing evaluation* in your essays too. This means taking every opportunity to reflect critically on the information that you are presenting to the reader. The annotated example at the end of this chapter (page 69) includes plenty of ongoing evaluation — review it carefully.

Making use of the specialised concepts

Exactly what are the elements of a good 'agree, but...' or 'disagree, but...' judgement? There are many approaches you could take, depending on the essay title in question. However, essay mark schemes give limited guidance as to *precisely* what is expected of a good answer. This is because every student may arrive at his or her own unique judgement. One important rule to remember though is that a good conclusion to a geography essay is one in which the writer is clearly 'thinking like a geographer'.

'Thinking like a geographer' is a phrase that is used widely in geography textbooks and examination mark schemes, but what does it really mean? Often, it indicates that the writer:

- makes reference to *geographical concepts* when concluding or as part of an ongoing assessment or evaluation — such as interconnections, interdependency, resilience or sustainability
- acknowledges that any conclusion may only be partial because the answer is dependent on the *spatial or temporal scale* of study — for example, a rebranding scheme may be successful in the short term yet may 'run out of steam' in the longer term
- shows awareness that people in different places may have *varying perspectives* on whether the statement is true or not, and that there may therefore be no definitively right or wrong answer to the question
- makes use of new theories, ideas or real-world examples, which throw into question previously accepted explanations

The specialised A-level Geography concepts are listed on page 10. It is worth referring back to these whenever you need to plan an essay or write a conclusion. It could be that one or more of these concepts might be useful. Table 4.3 reviews all of the suggested essay titles that have featured in this chapter so far. In each case, possible concluding points are shown that indicate the writer is 'thinking like a geographer'.

Table 4.3 Suggested ways of 'thinking like a geographer' that might feature in the concluding paragraph of an evaluative essay

Essay title	Possible ways of concluding that indicate the writer is 'thinking like a geographer'
Discuss the view that barriers to globalisation are increasing over time.	Currently evidence can be found that both supports and opposes this view at the *national* scale (such as 'Brexit'). At a *global* scale and in the *longer term*, however, barriers to globalisation will probably continue to decrease rather than increase overall because of further technological progress and the power and growth of the largest MNCs such as Apple, Amazon and Google.
To what extent are processes of deposition beneficial for human activity in coastal environments?	Both benefits and costs result from coastal deposition processes. In certain *local places*, such as estuaries, long-term deposition is not beneficial on account of problems created for shipping. However, the majority of depositional environments, including most of the UK's beaches, can offer great social and economic benefits, provided they are managed *sustainably*.
Assess the economic impact of international migration for different countries.	Both positive and negative economic impacts of migration can be experienced by source and host countries alike. However, these impacts are experienced by some places and cities more than others at the *local scale*, which can make it hard to generalise about *national-scale* impacts. Finally, the *longer-term* economic impacts of *greater interdependency* between source and host countries — such as Poland and the UK — could be very difficult to assess reliably.
Evaluate the economic impact of new flows of money and investment on different places you have studied.	On balance, the economic impact of flows of money is positive for many places. However, *inequality* can result when benefits are spread unevenly among local populations. Some people suffer from rising rents in gentrified places and *from their perspective*, impacts are not always positive. If housing becomes too expensive, and key workers such as nurses and teachers migrate elsewhere, then further growth of economically 'over-heated' places will become *unsustainable*.
'Vulnerability to tectonic hazards has risen over time.' Discuss this view.	In general, this statement is untrue *globally* if vulnerability is interpreted as loss of life. Fewer people now die as a result of earthquakes, volcanoes or tsunamis than in the past, because most countries have more effective warning systems than they used to. However, far more people today are *at risk* of being affected by tectonic hazards to some degree than in the past due to population growth and rising affluence, especially in megacities close to plate boundaries. In many of these places, it would be true to say that vulnerability *at the local scale* has indeed risen.
To what extent do you agree with the view that global governance of Antarctica has been successful?	Many people view the global governance of Antarctica as a success story overall. A high level of international cooperation and agreement has resulted in the continent being protected as a wilderness. However, *from the perspective* of those who believe that this important global commons should be entirely protected from human activity, the growth of Antarctic tourism might be viewed as a governance failure. The opposing view to this is that it is unrealistic for anywhere to be kept entirely 'off limits' *in an interconnected and globalised world*.

Including diagrams in an essay

It is perfectly acceptable to use diagrams, including sketch maps, system diagrams or landform illustrations, as part of your answer to an essay. In general, students are most likely to take this approach when answering physical geography questions. The use of diagrams is, in theory, to be encouraged in both physical and human geography because:

- it may take less time to show the features of a landform by drawing it than by describing it
- some theories and systems can be described and explained more quickly using a diagram than text
- an illustration may add clarity (especially if a complex idea or feature is difficult to convey using words)
- visual material helps to break up a body of text and may make the essay more enjoyable for your teacher or examiner to read

However, some types of diagram do not add any value at all to an essay. Moreover, when writing an essay under timed conditions in an examination it can be unwise to waste too much time drawing diagrams if it prevents you actually writing down all the points you want to.

- Do not draw a large map of the world with an arrow pointing towards the country you are using as a case study. Assume that the examiner already knows where Bangladesh is.
- Do not include a diagram of a landform whose descriptive labels (such as 'steep cliff face') use exactly the same words that you have already included in an earlier paragraph of writing. All you have done is repeat yourself using mixed media.
- Do not add cartoons of migrants jumping out of aeroplanes, or people walking into a McDonald's restaurant in China.

Ideally, the diagrams should convey information that is (i) important and (ii) not included elsewhere in your essay. There should also be material that can be awarded AO2 credit in addition to points that gain AO1 credit.

The difference between...

Diagram B would gain more credit than Diagram A as part of an answer to the evaluative essay question:

Assess the role of wave action in the development of two or more coastal landforms.

Can you work out why? Can you identify potential AO1 and AO2 credit here?

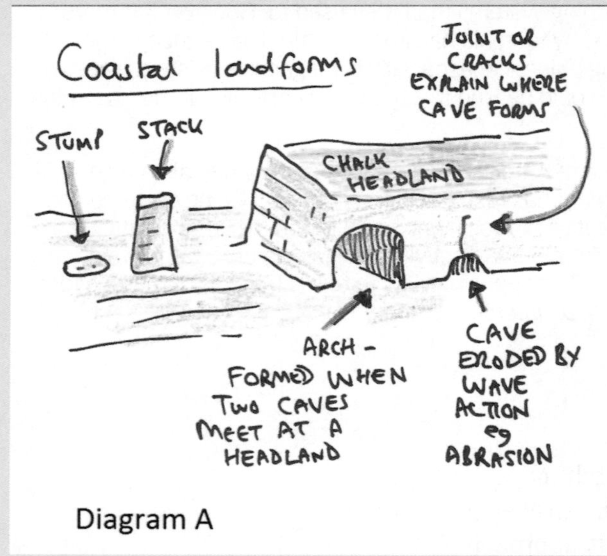

Diagram A

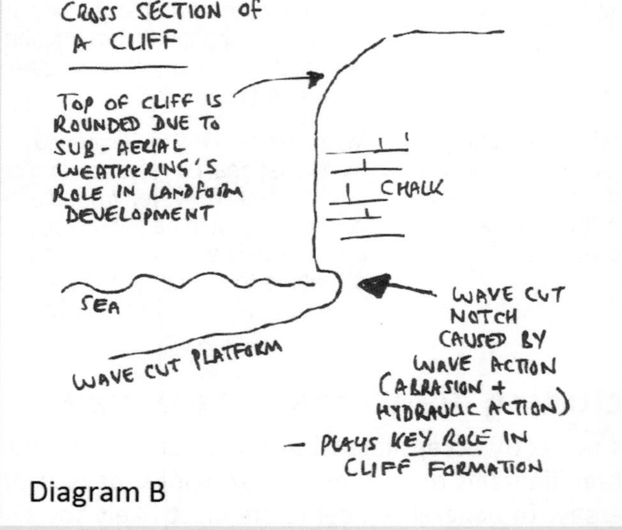

Diagram B

Application to the exam

Writing an essay under examination conditions is a very different experience from carefully crafting one at home. You can spend as long as you like (within reason) when planning and writing a homework essay; all the facts you need can be looked up online or in textbooks. In contrast, the examination requires you to not only recall memorised facts, but also to build an 'on-the-spot' evaluation of this material.

By following the principles and guidelines for essay writing outlined in this chapter, you will hopefully be well prepared to plan and

Exam tip

The best way to succeed at running a marathon is to do plenty of training beforehand. Similarly, if you want to succeed in writing essays under examined conditions, then you need to practise. There is nothing stopping you downloading specimen examination papers (available online) and writing essays under timed conditions at home.

write evaluative essays under timed conditions once your final examinations begin. The remainder of this chapter offers some final advice about how best to apply your essay-writing skills in the 'live' context of the examination itself.

The difference between...

Failing to focus on the question that has been set can prevent a student attaining a pass grade when writing under timed conditions (let alone an A-grade!). Try to spot the difference in quality between the following extracts taken from essays written under timed conditions in an examination, in response to the following question:

'Every country will eventually lose its distinctive national identity as a result of globalisation.' How far do you agree with this statement?

Extract from a lower-grade answer	Extract from a high-grade answer
TNCs such as McDonald's have spread around the world on account of globalisation. Their power is enormous and they can be found in any city, although often they change the ingredients they use, such as the McArabia burger, which is available in Middle Eastern countries. This is called glocalisation. Often, the menus are written in English, and it is estimated that because of this, 3000 world languages will soon be lost out of about 6000.	TNCs like McDonald's can be found in almost every country in the world, with the effect that cities begin to look the same wherever you are. This is a way in which national identity can be threatened, as places no longer look as individual. The spread of English as part of cultural imperialism is another way in which nations lose their sense of identity. Some languages become lost and so a key part of a country's culture is no longer as distinctive.
This is a knowledgeable student, who has correctly applied two relevant themes (TNCs and the dominance of English). However, there is no reference at all to the concept of 'national identity' here. An examiner is likely to make comments such as 'point unfinished' or 'not answering the question'. The most likely outcome is a low pass grade.	*There is no more evidence of geographical knowledge here than in the previous response. However, it is being applied in a more direct and explicit manner. Clear links are established, with the importance of language and the cultural landscape as distinctive signifiers of national identity. This has every chance of reaching the highest mark band available.*

Balancing AO1 and AO2 in an essay plan

A strategy that consists of 'start writing and hope for the best' is unlikely to be a winning one. Making an essay plan before starting writing is *essential* if you are striving to reach the highest grades. Surrendering a few precious minutes of time in order to develop a winning plan is your best bet for achieving an A or A* grade. You will doubtless know the saying that it is 'quality not quantity' that matters; this is very much true of essay-writing under timed conditions. Feedback offered to teachers by senior examiners in their annual subject examination reports often makes mention of the large numbers of students who paid insufficient attention to the wording of essay questions and instead adopted a 'write all I know' approach (and receiving a low grade as a result).

An essay plan need not take very long to write, especially if you are well versed in the 'rules of the game'. Figure 4.2 shows an example of an essay plan that has taken no more than 2–3 minutes to devise and offers a perfect 'roadmap' to success for this particular title. In particular, note the following:

- Only single words or short phrases have been used — it is important not to waste time writing an essay plan that is virtually a miniature version of the essay itself.

- The plan balances factual knowledge (AO1) with arguments (AO2) exactly in line with the requirements of the assessment.
- At the planning stage, the student has already anticipated what the conclusion is likely to be and has identified nuanced points that will need to be highlighted when concluding (these have been underlined).
- Key words from the title, along with the command word itself, feature throughout the plan. As a result, the completed essay is likely to be very tightly focused.

Assess the role of vegetation in the water cycle

Store? Flow?

Different biomes (e.g. seasonal forest)

Local or global? Arid or humid?

Intro — Mention STORE, FLOW, BIOMES, SCALE

Para 1 — Role as STORE (rainforest, grassland, desert)

Assess — varies in importance spatially and seasonally

Para 2 — Role in creating/controlling FLOWS

Transpiration, interception, overland flow

Assess — KEY influence

Para 3 — Role at local level — DEFORESTATION + floods in CUMBRIA

Conclusion — Plays key role locally, globally, seasonally

Figure 4.2 A sample essay plan

Tackling 'assess' and 'evaluate' command terms

Assess the importance of governance in the successful management of tectonic mega-disasters. **(12)**

(Edexcel sample assessment materials)

Evaluate the relative importance of strategies used by powerful countries to maintain global influence. **(20)**

(Eduqas exam-style question)

As we have already seen, 'assess' and 'evaluate' essay titles often make use of a word such as 'role', 'importance' or 'success'. Planning an effective response to a question beginning with one of these command words may involve thinking critically (and quickly) about any *underlying assumptions* that feature in the essay title. Underlying assumptions are hidden dimensions of the question which a good geographer might be expected to offer comment on, such as:

- The **timescale** of any strategies or changes that you will write about. In the Eduqas example above, do we assume that the focus is meant to be present-day strategies, or might we look at strategies used in the past? Part of the evaluation might involve looking at how the relative importance of different soft and hard power strategies has *changed over time*.
- The **places** and **scales** that might be used as examples. In the Edexcel example, an assessment might be offered of the

Exam tip

Do not describe your case studies and examples in excessive levels of detail. Instead, be selective and explain how particular aspects of your examples are relevant to the point you are making and the extent to which they prove a line of argument.

Exam tip

Make sure that you know exactly what the allocation of AO1 and AO2 marks will be in the essays you have to answer. This gives you a good idea of exactly how discursive your writing needs to be.

importance of local and national governance; in the case of the USA, both state and federal government play important roles in the governance of Californian earthquakes, for example.

- The **hidden dimensions** of key words included in the title. In the Edexcel example, 'successful' could be interpreted in varying ways (lives lost, financial costs, etc.). In the Eduqas example, 'global influence' has a range of potential meanings.

A good general strategy to adopt when planning an 'assess' or 'evaluate' essay might be to:

- underline the words that provide the **main focus** of the essay (such as 'successful management' or 'global influence' in the examples above), then quickly write a brief list of all the case studies you know about that could be relevant
- ring all the words and phrases in the essay title where an **underlying assumption** needs to be explored; think critically about contrasting ways of approaching these assumptions and identify which of your case studies will be the most interesting and potentially rewarding to use

Figure 4.3 shows the thought process you might follow when planning an answer to the examination essay question:

Assess the benefits of a shrinking world for different people and places.

A proper assessment should look at any **underlying assumptions** embedded in the statement. For example, don't just write an essay about the benefits experienced by 'most people' or 'people in general'.

'Shrinking world' is a fairly straightforward focus for the essay. However, at the planning stage don't forget that it could be assessed in contrasting ways. Importantly, we could distinguish between the shrinking effect brought by **transport** (container shipping, air flights) and the shrinking effect brought by **ICT** (Skype, social networks).

Assess the benefits of a shrinking world for different people and places.

What is meant by 'benefits'? One approach might be to think of **economic**, **social**, **cultural** and **political** benefits (a framework that is often applied to the analysis of globalisation or development: it is useful to compartmentalise 'big ideas').

What does 'different people and places' mean? There are simple distinctions (**HICs/LICs**) to consider. A more complex national-scale analysis might include **emerging economies**. Scale could be important here too: do all **urban and rural** societies in the same country have the same level of exposure to the 'shrinking world'?

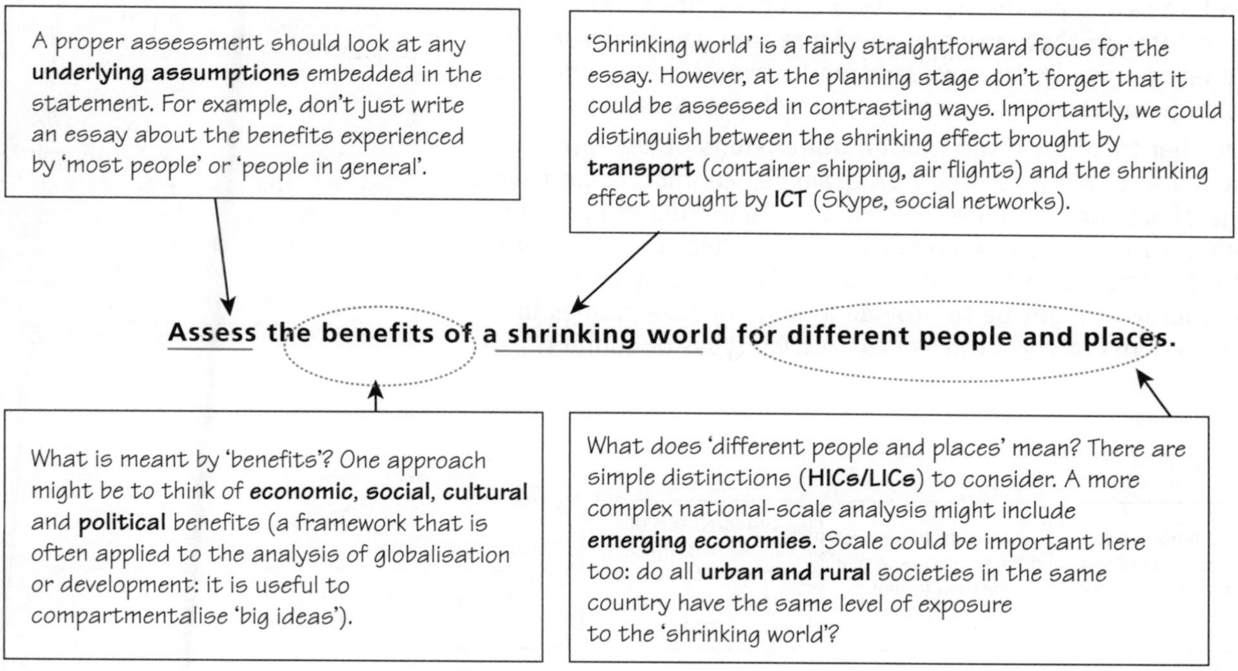

Figure 4.3 Analysing a question before planning your essay

Take it further

Now you have had a chance to look at Figure 4.3, devise a brief essay plan that uses a series of case studies to incorporate all the important themes that need to be covered. Keep your plan short (no more than 50–100 words, using just simple words or short phrases).

Tackling 'discuss' and 'to what extent' command terms

'Social factors are the most important influences responsible for gender inequalities.' Discuss. (16)

(OCR sample assessment materials)

'Conflict often arises when people who live in a place try to resist changes that appear to have been forced upon them by organisations, groups and individuals from outside that place.' To what extent does this statement apply to one or more places that you have studied? (20)

(AQA sample assessment materials)

'Discuss', 'to what extent' and 'how far do you agree' are command words and phrases used by all exam boards except Edexcel. These commands require you to construct a balanced response that gives roughly equal weighting to both sides of an argument.

- The first step when writing an essay plan is to decide which of the case studies and examples you have learned about can be best used in support of the two different sides of the argument.
- Using the example of the OCR essay above, you may have studied in depth a number of social factors that influence gender inequalities (for example education, inheritance laws, religion and other cultural traditions). At the planning stage you may therefore take a pragmatic decision to only write at length about two or three of these in order to ensure that you have sufficient time left to give *equal* consideration to two or three non-social factors (such as possible economic or political reasons).
- Remember to 'unpack' the **hidden dimensions** of key words included in the title. In the AQA example, 'conflict' could have a range of meanings. Figure 4.4 shows the spectrum of possible meanings for 'conflict'. A good essay may reflect critically on this and other 'hidden dimensions' included in the essay title. The best approach might be to provide a range of case studies in order to deliberately showcase *contrasting types* of conflict.

Exam tip

A really good answer to the OCR question about gender inequalities might consider ways in which social factors could be the most important 'root cause' overall *because of the way they also drive political and economic factors*. This is because political and economic decision making is usually shaped by a society's social norms. This argument could provide the basis for a powerful conclusion.

Social tension and disagreements over a proposed change to a place	Legal conflicts between different parties, e.g. in relation to new building plans	Isolated acts of vandalism or property damage; violent threats made using social media	Violent street protests or demonstrations; physical assault

Figure 4.4 A spectrum of possible 'conflicts' between different individuals or groups of people

Arriving at a conclusion in the examination

As we have seen, the final ingredient for an outstanding answer to an evaluative essay is a conclusion, which includes a *final judgement*. This means that the evaluation goes one stage further than merely recognising that there are two sides to the debate, or multiple ways of viewing the importance (or success) of something. Depending on the exact wording of the question that has been set, the final judgement may:

- say whether the statement under discussion is, in general, true or false
- provide a summative statement about the importance (or role, or contribution) of whatever has been the focus of the essay

Doing this for homework is one thing; managing it under timed conditions in the examination is an even higher order of challenge. The best advice to give, as ever, is: get plenty of practice.

The essay mark bands that appear in A-level Geography mark schemes are written in ways that suggest there is a very high expectation here, even though the essay must be written in a hurry under timed conditions. Top-level mark band phrases include the following:

- 'Detailed *evaluative* conclusion that is *rational* and *firmly based on knowledge* and understanding which is applied to the *context* of the question.' (AQA)
- 'Applies knowledge and understanding of geographical information/ideas to come to a *rational, substantiated* conclusion, fully supported by a *balanced* argument that is drawn together *coherently*.' (Edexcel)
- 'A detailed and *substantiated* evaluation that offers *secure judgements* leading to *rational* conclusions that are *evidence-based*.' (OCR)
- 'Applies knowledge and understanding to produce a thorough and *coherent* evaluation that is supported by *evidence*.' (WJEC/Eduqas)

What does all this mean in practice? How do the examiners know when they are looking at a 'secure judgement' or a 'rational' and 'coherent' conclusion? Table 4.4 provides contrasting conclusions for two essays written on the topic of global systems. In each case you will see that the higher-scoring essay has a conclusion that may be *substantiated* (refers back to key evidence) and/or is phrased in a *more careful and thoughtful way* (which might lead an examiner to view the work as being more 'rational' or 'secure').

Table 4.4 Sample conclusions for two essays

Essay title	Basic conclusion (Student 1)	Secure conclusion (Student 2)
'Globalisation is mainly driven by the actions of transnational corporations (TNCs).' **How far do you agree with this statement?**	Having looked at both sides of the question, on balance I believe that, nowadays, TNCs are the most powerful force driving globalisation. They have more money than many poor countries, and can make their products profitable in almost any country by using glocalisation, as this essay has shown. However, technology is important too, especially the internet, so it is not just TNCs that are important in making globalisation happen.	In conclusion, TNCs act alongside other forces. I certainly do not agree that TNCs drive political globalisation — this is more likely the work of the UN, EU and IMF. Furthermore, powerful governments like the USA have driven geopolitical changes that make it easier for their own TNCs to promote economic globalisation. Finally, technology *in conjunction with* TNCs drives social globalisation (e.g. Facebook). So while the statement may be largely true for economic globalisation, I have shown that different forces are driving other aspects of globalisation.

Essay title	Basic conclusion (Student 1)	Secure conclusion (Student 2)
Evaluate the impact of globalisation on the physical environment.	So, all in all, environmental damage is worse in poor countries such as Ghana where e-waste is sent and where polluting factories are built. Rich countries do not suffer from this pollution as much anymore, so I do not agree that environmental damage is experienced everywhere. But overall the impacts are mostly negative and the worst impacts can be highly damaging. Climate change is the worst impact of all, as I have shown in this essay.	In conclusion, everywhere is subject to some harm at a *global scale*, due to the planet-wide impacts of global warming and oceanic pollution. However, even more extreme localised impacts sometimes take place too, at a *local scale*, due to weak local governance attracting polluting TNCs, such as Ghana's government allowing harmful and damaging e-waste processing. Finally, we must remember that *improvements* do sometimes take place as a result of the work of campaigning global NGOs like Greenpeace.

The difference between...

Ingredients of a basic (or 'assertive') conclusion	Ingredients of a secure (and persuasive) conclusion
• Asserts that the statement is either mostly true or false (no further qualification of this view is offered). • Concludes the discussion by saying that it is difficult to decide whether the statement is true or false overall (even though the evidence used in the essay may strongly suggest otherwise). • Writes in a way that suggests no attention whatsoever has been paid to any of the underlying assumptions in the title (for example, there is no recognition that the truth of the statement under discussion depends on what spatial or temporal scale is adopted when studying the issue). • Begins the conclusion with an outdated or juvenile expression such as 'So, all in all…' or 'Nowadays…'.	• Acknowledges explicitly that any conclusion reached is likely to be *partial* because there are so many different ways of thinking about the issue, or so many *perspectives* to consider. • Refers back to the most *significant* facts that have featured in the main body of the essay (in support of an evidence-based final judgement). • Makes reference to specialist geographical *theories*, terms or ideas in order to add 'weight' to the final judgement. • Establishes *connections* between different ideas and themes that are featured in the essay. For example, the conclusion to an essay discussing 'the relative importance of different factors that affect hydrological flows' might recognise that climate is the most important factor because it *in turn* influences other factors, such as vegetation and soil type.

Activity

What is your own view on **sustainable development**? In almost any A-level Geography paper you will find an evaluative essay question that asks about sustainability in one form or other. Are the place-making schemes you have studied sustainable? To what extent is global development sustainable? Can coasts be managed sustainably?

More generally, how far do you agree that the Earth itself has a sustainable future?

- It is worth taking time to work out what *your own* view on this critical issue is, and what the evidence or arguments are that lead you towards whatever judgement you make.
- It could be that you are a *pessimist* because all recent data showing population growth, rising affluence and resource consumption trends indicate that current actions to support sustainable development are 'too little, too late'.
- Equally, you may be an *optimist* who sees the future as bright because of the power of technology to bring about rapid and positive change (there are plenty of examples from history that can be used as evidence to support this view).
- Whatever you believe personally, the important thing is to be able to persuade other people that your view is rational, intellectually secure, grounded in evidence and worth taking seriously.

Annotated example

Evaluate the role played by local people in bringing economic changes to urban places.

It is important at the planning stage to think carefully about:

- what is meant by local (and thus 'non-local') people
- different kinds of economic change (e.g. the kind of employment that is available or the way land is used; also, the degree of change over varying timescales)
- the scale and context for change (neighbourhoods or entire cities could be discussed; examples can be drawn from the UK or other countries)

In the answer that follows, you will observe that these underlying assumptions have been carefully 'unpacked' in the introduction to help prepare for an effective debate and conclusion.

From the outset, this essay is tightly focused on economic change, which suggests that it will score highly according to A01 mark scheme criteria.

Economic changes occurring over time in urban places include changes in the kind of work that is done there (such as manufacturing, services or quaternary employment) and land use (factories, shopping centres or research facilities). Another changing economic characteristic is the economic profile of the population (e.g. the proportion of people in low-income or high-income groups). This essay will evaluate the role of local businesses, governments and community groups in bringing change to local urban neighbourhoods and, at a larger scale, entire cities. The role of external players — including foreign governments, foreign investment and flows of migrants — will also be discussed.

This introductory paragraph sets the stage for a discussion that should score well according to A02 mark scheme criteria. The writer has established that an important counter-argument exists.

The regeneration of Salford Quays in the Manchester region provides a good example of the way positive economic change can be driven by local players acting together. During the 1990s, the local water company United Utilities invested billions of pounds in cleaning up the Manchester Ship Canal and improving water quality, which was seen as essential for economic regeneration of the area. Local experts from Manchester University played an important role by designing the equipment used to help re-oxygenate water. The local property development company Peel Holdings invested large amounts of money in the area, helping to redevelop waterside properties. The result of all of this was a complete economic transformation of Salford Quays, from a derelict industrial area into a thriving hub for services and leisure. None of this would have occurred without the involvement of local players. However, the UK government did play an important role too by 'getting the ball rolling'. Also,

This paragraph is very well structured according to the 'PEEL' principle. This means that a relevant point (P) is made and then explained (E). Evidence (E) is provided and clearly linked (L) back to the question. Throughout, very strong links are made with the essay title — look how many times the words 'local' and 'economic' are used.

The last two sentences provide ongoing evaluation. The keyword 'however' is used to introduce an element of critical thinking to the case study by briefly pointing out that while the writer may strongly agree that changes in Salford Quays were locally driven, they were not *entirely* locally driven.

some economic changes were driven by non-local external investment, such as the BBC's decision to relocate from London to Salford.

Elsewhere in the UK, there are many other examples of urban regeneration led by local business partnerships working alongside local government. During the last 20 years, new flagship developments have appeared in many British cities. These include the new Bullring in Birmingham, Cardiff's Millennium Centre and Newcastle's £70 million Sage Gateshead Centre, which was completed in 2004. These flagship schemes have played a key role in changing people's perception of these cities and have encouraged further economic changes, including the growth of new retail and office spaces, leading to changing employment structures. In the past many people in Birmingham worked in heavy industry but now the majority work in service industries. In the case of Birmingham, recent changes owe a great deal to the work of the Greater Birmingham and Solihull Local Enterprise Partnership, which is an alliance of local people representing businesses, local authorities and universities. Once again though, none of this might have happened without UK government establishing Local Enterprise strategies in the first place.

In contrast, there are some examples of economic change in urban places that are far more obviously driven by *outside* forces rather than by local people. The gentrification of urban neighbourhoods is an obvious example of this. Gentrification is a process of economic and social change, which involves affluent people moving into low-income or derelict neighbourhoods. A 'snowballing' effect then takes place that involves the arrival of more high-income people and investment by businesses once the local area's image begins to improve (although older local residents may not always view the changes as an improvement). Gentrification in Hoxton (London) has totally transformed the area's economic characteristics. House prices are 10 times higher than 20 years ago and the area is now full of young media and technology professionals who have come from other places.

The use of data here means that the essay is likely to score well according to A01 criteria. You should always try to include some factual evidence in each essay paragraph.

As in the previous paragraph, there is ongoing evaluation here, which will help to maximise A02 credit.

It is good practice to define any concept or terminology that is introduced into the essay in order to give your essay an 'expert voice'.

This is important detailed evidence for economic change, which means that A01 credit is more likely to be gained.

This is a good 'connective' phrase to use, which helps tie the paragraphs of the essay together into a coherent, well-structured account.

Finally, it is important to note that all of the economic changes discussed in this essay are part of long-term changes driven by global forces. It can even be argued that the most important influence on changing urban places in the UK has been the rise of emerging economies (such as South Korea and, more recently, China). The industrialisation of these countries is a big reason why deindustrialisation occurred in British cities from the 1960s onwards. The actions of local business partnerships and attempts to regenerate urban areas are therefore a reaction to externally driven changes that were seen at first as very undesirable by local people because of the unemployment created.

This is a crucial statement because it potentially 'wins' the argument that local people do *not* have the most important role in bringing change to places if we think about 'the bigger picture'.

In conclusion, both local people and external forces have played important roles in bringing economic changes to urban places in the UK. New city flagship developments have been managed by local players, and it is safe to say that local people often have the most important role in deciding exactly *how* regeneration is carried out at the local level. However, regeneration relies on external flows of money and people too, and arguably would not even be needed if not for deindustrialisation caused by economic changes happening in other places. Taking a long view, outside forces and not local people play the most important role because they are the whole reason why economic change in cities has become necessary.

A substantial reflective conclusion is presented, which is well linked to the essay question. Overall, the writer has chosen to take a 'disagree, but...' stance. This is a well-justified and entirely defendable viewpoint given the contents of the essay. Overall, this essay would surely deserve full marks.

You should know

- There is a fundamental difference between descriptive and evaluative essay-writing styles.
- Make sure that you know what your own exam board's specific mark scheme requirements are for evaluative essay writing.
- Try to anticipate possible evaluative essay titles for the topics you study by looking carefully at how your specification is worded (using words and phrases like 'role' and 'the importance of').
- Writing an essay plan is essential for exam success; remember to include facts and argument and make sure you provide a balanced discussion that takes different points of view into account.
- Make sure that you know what the ingredients of a really good conclusion or final judgement are. Do not sit on the fence, but also remember not to dismiss some arguments out of hand. The best approach could be to conclude 'I agree, but...' or 'I disagree, but...'.

5 Making an outstanding independent investigation

Learning objectives

- To get to know the main elements of a high-quality independent investigation
- To follow a step-by-step guide to ways of maximising your independent investigation grade
- To know the common pitfalls to avoid during the planning and fieldwork stages
- To understand how to make your fieldwork analysis and presentation appear outstanding

Planning your investigation

As part of your A-level Geography course, you must complete a fieldwork-based **independent investigation** (also known as the non-examined assessment, or NEA).

- The independent investigation consists of a written report.
- It is marked by your teacher.
- Ideally, it should be between 3000 and 4000 words in length.
- The focus must be related recognisably to your A-level Geography specification.

Aiming for high quality

Producing a high-quality independent investigation is the single most important thing you can do to improve your overall chance of gaining an A-grade in geography. It is worth 20% of the A-level. Although this may not sound a lot, it can represent the difference between an E-grade (40–50% of total marks) and an A-grade (60–70%). Your written report will therefore play a crucial role in determining whether you reach the A-grade threshold.

In order to excel in the independent investigation, you need to have a clear vision at the outset of what you want to achieve:

- The best preparation is to make sure that you are genuinely interested in the topic you choose to focus on. This intellectual curiosity should promote greater engagement with background literature and deeper thinking about all stages of the enquiry process.

Exam tip

One of the first things to do is get hold of a copy of the mark bands used for your exam board's independent investigation. You must have a clear vision of what your written report needs to achieve.

→ When you are planning what to do (where to visit, what to measure, etc.) you need to keep thinking ahead about the later stages of the investigation (such as the use of data presentation techniques and the evaluation you will need to offer about the accuracy and reliability of what you did).

Table 5.1 offers a generic summary of the different stages of the independent investigation used by the English and Welsh exam boards.

Certain key criteria have been highlighted: when your teacher (or an external moderator) marks your report, they will pay close attention to whether you have met these goals. In order to justify giving you a higher mark than many other students, your teacher must be convinced that your report fits the marking criteria very closely.

Table 5.1 The stages of your independent investigation

Stage		What your written report must do
1	Context setting	• Identify background to the topic focus, including any conceptual framework or important theoretical background. • Reflect on **risks** (the risk assessment) and **ethical issues** arising from the delivery of the independent investigation. • Outline the case study area succinctly to provide context.
2	Field investigation methods	• Use a good range of appropriate fieldwork methods to observe, measure and record geographical phenomena (ideally both quantitative and qualitative data). • Demonstrate **critical thinking** by **justifying** the methods used (e.g. their viability and appropriateness for the research question(s) under investigation). • Clearly outline **your own independent contribution** to the design of the methodology.
3	Data presentation techniques	• Communicate **primary** and **secondary** data findings using a range of appropriate presentation techniques, ideally including spatial (mapping), graphical (charts) and representational (photographic and other visual) methods. • Reflect on the **appropriateness** of the techniques used, and the extent to which they allow a better-informed analysis of the data.
4	Analysis and interpretation	• Ideally make use of statistical methods and measurements. • Identify patterns, trends, **anomalies** and possible causal relationships. • **Justify** any evidence-based arguments derived from data analysis and interpretation, most likely by reflecting on the **significance** of the investigation's findings, and levels of **confidence** in what the data show.
5	Conclusions and presentation	• Conclude in a **well-evidenced** way using **all** of the investigation's findings. • **Refer back** to the theoretical or conceptual underpinning of the research, and the extent to which the findings correspond with established geographical knowledge or perhaps offer new insights. • Follow presentation rules well (e.g. the **bibliography**/referencing of secondary data).
6	Evaluation	• Provide a succinct reflection on **every stage** of the investigation and the extent to which strengths and limitations can be identified in what was done. • Offer a **comprehensive** overview of primary and secondary fieldwork methods and data, in terms of their **accuracy, reliability** and possible **bias**, and so reflect on the **extent to which** the conclusions can be relied upon. • Reflect on how the views and interests of different **stakeholders**, if relevant, may have influenced the investigation's findings. • Suggest ways in which a similar investigation could be improved upon or extended with further research.

The difference between...

High-quality independent investigation	Low-quality independent investigation
• The work has been driven by genuine curiosity about a curriculum topic (evidence for this may consist of an impressive reading list). • Meticulous fieldwork is analysed rigorously in ways that are mindful of a wider geographical 'big picture'. • There is critical thinking and reflection about every stage of the research process.	• The work is uninspiring and has been based entirely upon a small area of textbook content with little or no evidence of any broader reading or critical thinking about the topic. • There is no evidence that the fieldwork was carried out with particular care or thought. • A narrow analysis is full of missed opportunities for thinking geographically.

Finding a focus and question

Coming up with your own geographical idea or focus can be challenging.

→ Your teacher may encourage you to work entirely independently. Essentially, you will be on a 'solo mission' with a topic and case study area of your own choosing.

→ Alternatively, your entire class may be expected to produce independent investigations that draw on a shared field-trip experience.

Table 5.2 summarises these different approaches and the important steps you will need to take in order to maximise your chances of gaining a high grade.

Table 5.2 Different approaches to your independent investigation

Group field-trip-derived independent investigation	'Solo mission' independent investigation	Combined approach independent investigation
• A school field trip could serve as the basis for your own independent investigation. • But you *must* demonstrate that you *personally* had an active role in designing the fieldwork: you cannot just 'follow orders'. • This means that you need to plan ahead carefully. *Well before the fieldwork visit begins*, make sure you have (i) decided what your own research focus will be and (ii) thought of ways to *adapt* or *modify* part of the work that your class will be carrying out. • Failure to take these steps could reduce the marks you can be awarded.	• You produce an independent investigation that is entirely separate from any whole-class field trip(s). • For example, your class might visit a field studies centre for four days in April (Year 12) to carry out coastal fieldwork. Having done this, you decide to carry out your own independent investigation researching infiltration rates in local parks at home during the summer holiday. • You might be able to adapt fieldwork methods from the class field trip. • This is a highly independent approach to take.	• It is possible to combine approaches. For example, your class might visit a field studies centre for four days in September (Year 13) to carry out fieldwork exploring the cultural diversity of Town A. The entire class participates in the collection of primary data for Town A. • You then carry out a follow-up study at a second location (Town B) where you live. You refine the fieldwork methodology used for Town A, and collect new data for Town B. This could be done over October half-term. • Your completed study would give a unique comparison of Towns A and B.

Aims, questions and hypotheses

Once you have identified a broad topic for your independent investigation (such as coastal landscapes or the water cycle), you need to develop specific aims, questions and/or hypotheses.

- An **aim** is a statement of what your geographical enquiry wants to achieve. A well thought out enquiry aim will be geographically sound and achievable. For example: 'My aim is to investigate how Ainsdale beach is affected by longshore drift.'
- Key research **question(s)** are used to frame the enquiry. For example: 'How do beach profiles vary along the Ainsdale shoreline?' One question may serve as the overall enquiry title; or sub-questions can be used to sub-divide your work. Be careful not to devise *too many* sub-questions (three or four is a good number).
- A **hypothesis** is a statement whose accuracy can be tested objectively using scientific methodology. For example: 'People using Ainsdale beach for recreation are mainly from the local area.'

Using the specialised concepts

By now you should be familiar with the marking criteria for the independent investigation used by your exam board. You may have seen how credit is awarded for 'sophisticated', 'coherent' or 'reasoned' writing, especially when it comes to the final evaluation of the completed work.

- If you hope to meet these targets and access the highest mark bands, then you must think carefully *from the outset* about the opportunities for *sophisticated and reflective geographical thinking* that your independent investigation title is going to provide.
- For this reason, students are sometimes advised to make use of the **specialised concepts** for geography, such as risk, resilience, threshold and globalisation. These were introduced on page 10; you might consider including one of these concepts in the title of your investigation.
- For example, rather than asking the simple research question 'What proportion of shops in Woodford are British-owned?' you might instead ask 'To what extent is Woodford a globalised place?'
- The second title requires you to come up with an appropriate methodology. What would you measure? What data sources could serve as valid indicators of somewhere being a 'globalised' place?
- In turn, there are now greater opportunities for writing a thoughtful conclusion and evaluation — globalisation is a hard concept to pin down, so you can reflect afterwards on how valid and reliable you feel your data actually were.

Designing a rigorous methodology

'Rigorous' means being thorough and careful. For fieldwork, being rigorous involves:

- thinking properly upfront about the likely strengths and weaknesses of different possible ways of carrying out your fieldwork before identifying the best approach to take
- taking every possible step to record and analyse your data accurately

! Common pitfall

Some titles are too ambitious, for example: 'How successful is London as a global hub for migration?' You would be better off focusing on migration into a particular London neighbourhood.

Exam tip

Don't choose a question that is *too far removed* from geography. If you study English, then it could be tempting to make a study of how landscapes have been represented in classic novels. But be careful not to end up producing an English report instead of a geography report. The same risk holds for biological studies of the carbon cycle. If in doubt about how geographical your independent investigation will be, a good question to ask yourself is: *What maps might I be able to draw as part of the work, other than those used to establish the context?* If the answer is 'none', then it is possible that your planned work lacks sufficiently spatial elements.

Exam tip

Using high-quality background information and literature is essential for success. Specific marks are allocated for demonstrating that you have done this. Perhaps more importantly, background reading helps generate more interesting and challenging research questions, which in turn can inspire you to produce higher-quality work. Refer back to Chapter 1, which looks at the study skills needed to research and note-take effectively, including the production of a bibliography.

A rigorous methodology will therefore include a fully justified account of the methods used to collect different data, along with a rationale for why these data were deemed useful. For example, you might want to compare the environmental quality of two different urban places. This could involve measuring noise levels, traffic volume, litter, graffiti and the amount of green space. But if you only had time to focus on two of these, which would you choose? What rationale can you provide for why these two indicators are *most* important?

Sampling strategy

Your teacher and any external moderator will pay particular attention to the details of your sampling strategy. You should justify your choice, for example why you felt a stratified approach was more appropriate than a systematic approach.

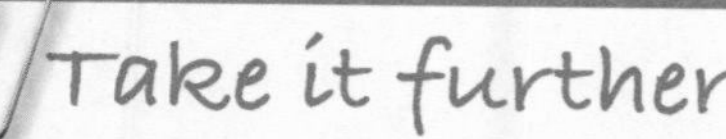

You can find out more about different sampling strategies at: www.rgs.org/schools/teaching-resources/sampling-techniques.

Exam tip

Many students will rely mainly on quantitative (numerical) data, for example primary data collected from a beach showing the range of pebble diameters. In contrast, other independent investigations will use more qualitative data (such as open-ended interviews). Try to use both types of data if you can. For example, a study of deficit within the water cycle may focus mainly on rainfall and soil moisture quantitative data. But you could also make use of old photographs and newspaper cartoons showing the famous drought of 1976 — then your work will include qualitative data too.

Making use of geospatial technology

Geospatial technology has become a great asset for students wanting to collect and manage data; it can support you through every stage of the enquiry process. This includes making use of online databases, geographical information systems (GIS) and phone apps when carrying out research. Because technology is advancing rapidly, especially in terms of the way smartphones can support geographical research, make a point of looking for the most recent resources that can offer advice on available fieldwork apps.

! Common pitfall

You cannot count all the pebbles on a beach or interview every person in a town. But how large should your sample be in order to provide a valid representation of reality? The answer is not 3, 4 or even 5. In most cases the minimum sample size is 10 (and 20 people or 100 pebbles would be better). If you are going to use a statistical test (like Spearman's rank) then make sure you have met the minimum sample size that the test allows.

Demonstrating independence: a summary

Figure 5.1 provides a useful summary of expectations for independence, both at the planning stage and later on in the process. In particular, note that:

- you must decide on your specific enquiry question(s) independently, but this decision can be framed by teacher input and guidance on the broad topic frame (*high level of independence*)
- you need to be involved in key decisions about what fieldwork procedures and data sources will be used in order to demonstrate active engagement with the process (*medium level of independence*)
- after the fieldwork has been carried out (working either with a group or alone on a 'solo mission'), you must analyse your evidence and draw conclusions independently (*high level of independence*)
- you are also expected to work alone to critically evaluate the validity of your evidence and the reliability of your methods (*high level of independence*)

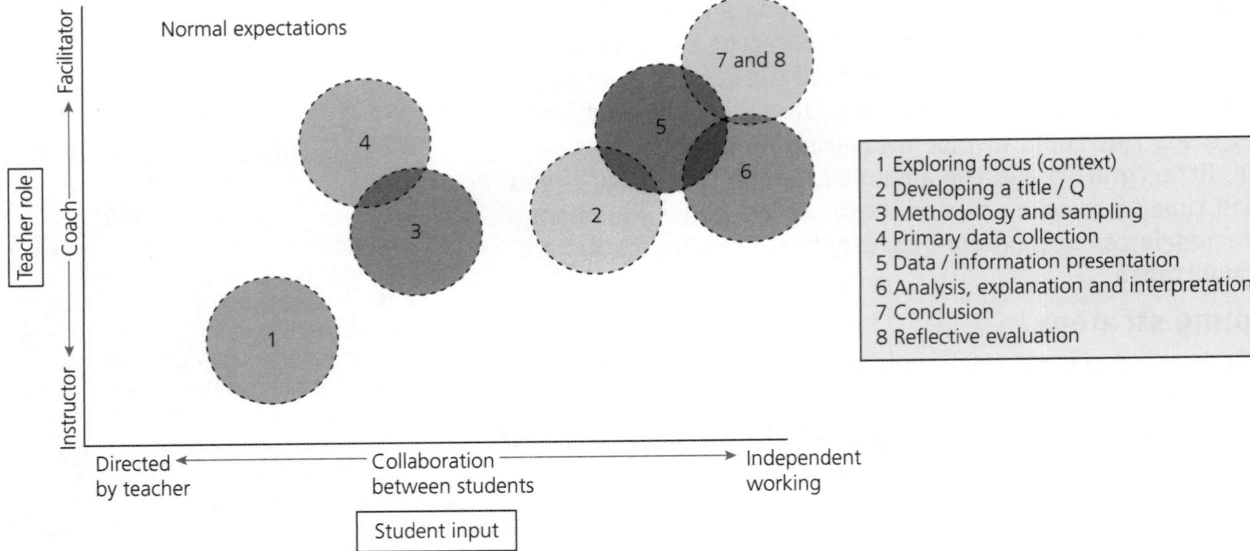

Figure 5.1 Normal expectations for the independent investigation
Source: Nick Lapthorne (ESC) and Becky Kitchen (GA)

Carrying out the investigation

To be successful, it is vitally important to plan carefully and stick to the deadlines you establish in advance.

- If you carry out your fieldwork as part of a school field trip, make your plans well ahead of arriving at the destination. Do not leave it until the night before you carry out your fieldwork to iron out the details of what you will be doing. This should have been done many weeks beforehand.
- If you plan to work alone, for example during the summer holidays, then you need a research plan that is realistic and achievable. You must anticipate any risks and ethical issues arising from the work you plan to undertake.

> **Exam tip**
>
> It could be tempting to carry out fieldwork in an exotic holiday location while travelling with family or friends, but there is virtue too in sticking closer to home (even if the topic and setting seem less exciting). The risk of working abroad is that it is impossible to return to the site if you forget to record something or want to carry out follow-up work. Carefully weigh up the 'wow' factor of foreign fieldwork against the many advantages of a home-based study area.

Thinking critically about your methods

When carrying out fieldwork, reflect on how you *interact* with the environment or other people because this could be affecting the validity and reliability of your data.

- Keep a research diary in which you write down any thoughts you have about this while working in the field.
- Anything you write down could prove useful later because you are expected to evaluate your methodology by writing critically about its strengths and weaknesses.

Are your data valid and reliable?

The validity of fieldwork data refers to the extent to which they are a sound and credible reflection of the real world (what you have set out to measure).

- A student who is asking people to comment on how they feel about a new shopping development should not ask 'leading' questions such as: 'Don't you think the new development is terrible?' This is a biased way of phrasing a question that could encourage people to answer 'yes'. Data showing that the majority of people dislike the shopping centre could be deemed invalid if the question asked has prompted people to answer in a particular way that may not reflect their true feelings.

→ Another example would be a student measuring soil infiltration rates who unthinkingly tramples and compresses the soil with her feet before pouring on water and measuring the time it takes to soak in. Simply by being there, she has modified the soil structure: the data she records are no longer a valid representation of the soil's infiltration capacity *in its natural and unaltered state.*

The reliability of fieldwork data refers to the extent to which information was recorded accurately (meaning that if someone else repeated your measurements or questions, they would get exactly the same findings).

→ Careless and inaccurate measurement of slope gradients would result in unreliable data (this is why you are allowed to collaborate with others to collect data — it may be essential to do so for the sake of reliability).

→ What steps did you take to maximise accuracy and reliability? A good independent investigation will always include some detail about this.

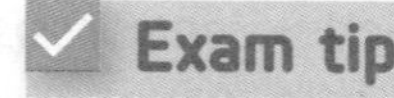

Exam tip

Think about possible validity and reliability issues affecting your own fieldwork and write about them in your report.

Table 5.3 provides a number of tips to follow when collecting certain types of data. Aim to collect a range of different data types *while also demonstrating the steps taken to ensure all data were valid and reliable.*

Table 5.3 Tips for collecting data

Surveys	• A well-designed questionnaire can generate large volumes of data, which can be manipulated graphically and subjected to statistical testing. Aim to perhaps combine **objective** measures, such as the distance people travel to visit tourist attractions, with people's **subjective** views about something, such as the desirability of a new development. • Qualitative data surveys can also be used in physical geography, for example investigating how people's **perceptions** of local flood risk may vary.
Interviews	• Semi-structured or unstructured interviews can take the form of a conversation lasting for perhaps half an hour. This is recorded with the consent of the interviewee (this is an important issue of **ethics**, which you should write about). • Not all topics lend themselves well to in-depth interviews. However, a study of the experiences of economic migrants or refugees might be well-suited to a small sample of in-depth interviews.
Sediment stores	• The measurement of the size of a beach store and/or its typical sediment size must be carried out in a **rigorous** way using tape measures, callipers and other equipment that is easy to handle. You may need to **justify** the use of teamwork.
Flows and movements	• Flows and movements can be studied within physical systems. Infiltration rates can be calculated using a stopwatch to time how long it takes a particular volume of water to drain from an infiltration ring or plastic pipe into the soil. But great care must be taken. • Human flows — such as traffic or pedestrians — can also be counted and rates calculated; this type of work may **justify** collaborative fieldwork.
Representations	• A wide range of visual data can be collected, such as photographs of urban graffiti, or the comments that people have posted about a place on a website such as TripAdvisor • You might even collect examples of old folk songs and music that have been written about a place, or use old newspapers to study how a football team and its fans have been portrayed in the past. But think carefully about your **sampling strategy**.
Patterns and distributions	• Patterns and distributions can be measured by sampling data at numerous different sites and recording the information on maps. • You can do this electronically by recording data using a **smartphone** (e.g. to take decibel readings in an urban area). The information can be automatically uploaded to a GIS base map and you can **reflect** on the advantages of using technology in this way.

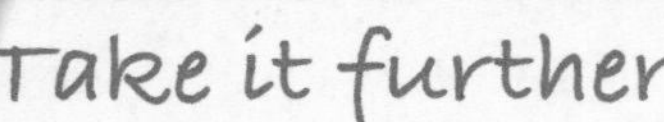

The types of data that can be collected, and the range of methods available, are widely covered by many textbooks and websites. In particular, the UK's Royal Geographical Society (RGS), Geographical Association (GA) and Field Studies Council (FSC) have produced quality materials that are, in most cases, freely available, and which you can use and adapt.

Completing the investigation

Devise a detailed timetable for writing up your investigation and stick to it.

- Year 13 is the most demanding of many people's entire school career.
- You are required to carry out your data analysis and evaluation independently and *without assistance from your geography teacher or any other teachers*. As a result, you may find that you get stuck occasionally, and need time to pause, think and reflect.
- Leaving the writing-up of your independent investigation until the last minute can also result in you failing to complete vitally important work in other subjects. It is far better to spread the work out over a longer period time.

Analysis and interpretation

Table 5.4 offers tips for the use of particular analytical and interpretive techniques. Make every effort to avoid common pitfalls that lead to mediocre instead of outstanding marks being awarded for your analysis.

Table 5.4 Tips for analysing your investigation

Analytical technique	Tips for using these techniques
Mapping	• Follow all the important conventions, such as including a scale and north arrow. • Be careful not to use too many colour classes, thereby over-complicating a map to the point where any patterns or distributions become harder — not easier — to see.
GIS	• GIS can help you analyse your data in innovative ways. Unlike paper mapping, GIS includes a database that you can continue to access at any point during your data analysis. • You can use filters to highlight locations that meet particular criteria (such as urban fieldwork locations where a pedestrian count higher than 50 was recorded). • It is well worth taking the time to find out more about this, for instance at: www.ordnancesurvey.co.uk/support/understanding-gis
Scatter graphs	• Be wary of always drawing a straight-line best-fit line through a scatter graph (see page 27). • For instance, the relationship between infiltration and precipitation is nonlinear, so you should *not* attempt to draw a straight line on a scatter graph showing precipitation and infiltration for rainstorms in a catchment.
Charts	• Try not to use overly simple bar and pie charts — after all, these are techniques that even primary schoolchildren can make use of. • Instead, think about better ways of *combining* data sets visually, for instance by using grouped or divided bars. • Rather than showing two separate 'before and after' pie charts, it is good practice to use two semicircles drawn back to back.
Annotated photographs	• Do not fill up space in your independent investigation with large numbers of photographs that lack any annotation or text. It is your job to add value to visual images. • Overlaying captions and text boxes, labels and arrows on photographs is relatively easy to do using a program like Microsoft Word.

Analytical technique	Tips for using these techniques
Less well-used graphical techniques	• Your exam board's specification contains a long list of different cartographic and graphical techniques, in addition to possible statistical calculation techniques. • Familiarise yourself with these, and search online for any you are unfamiliar with (an image search will be useful if you want to see examples of a Lorenz curve or logarithmic graph, for instance). • Figure 5.2 shows a range of possible techniques that you may not have considered using before. This may provide you with inspiration for interesting ways to present your own data that are relatively less widely used, and will help your independent investigation stand out from the crowd.
Qualitative data	• Qualitative data can be a challenge for data presentation. Many academics publish work using qualitative data which reads more or less like a newspaper article (i.e. a long essay containing quotations from people who have been interviewed). In order to meet the marking criteria for data presentation you should probably attempt to *visualise* any quotations from interviews. • For instance, you could overlay speech bubbles on an aerial photograph of the place you have been studying.
Chi-squared and Spearman tests	• For both of these tests, be careful to perform the calculation accurately and to communicate your findings effectively (see page 25). • You may decide to include an extract from a critical values table in your independent investigation, possibly as an appendix.

Finally, make sure you are familiar with all of the rules and instructions that need to be followed as part of the submission process. For example, the WJEC and Eduqas specifications require students to produce a word-processed report in Arial, Calibri or Times New Roman (11 point) font, with line spacing set at 1.5. All pages must be numbered. Read your specification carefully in order to find out if there are any procedures that you need to follow.

Exam tip

It is a good idea to make use of statistical tests such as chi-squared and Spearman rank correlation, provided it is appropriate to do so and your sample sizes are large enough to justify this (you would not carry out a Spearman rank correlation test for just four or five pairs of data, for example). It is one thing to be able to complete the statistical calculation; it is another to provide a meaningful account of what the test result shows (including testing at different confidence levels). See also pages 25–26.

Concluding and evaluating your independent investigation

The time will finally arrive when you need to write your report's conclusion and evaluation. There are several important rules to follow:

- A sound conclusion always refers back to the original fieldwork question, is based on the evidence collected, and is consistent with the results and analysis.
- Trends, spatial patterns and any anomalies found are identified, linked and discussed comprehensively.
- All concluding material should link back clearly to the data and any graphical or statistical treatment. Anomalies should be explained, not simply ignored or ascribed to some form of observer error.
- Where appropriate, conclusions may examine different perspectives in relation to the specific fieldwork question and the established geographical theory and context.

You must also offer a sensible evaluation of your enquiry, with valid suggestions for improvements. These suggestions may relate to the choice of case study context, the methods and equipment used, or any other aspect of the research process. Suggesting that more data could have been collected or greater care taken will not impress your reader, however. A good critical evaluation will dig deeper than that. For instance, with hindsight, how satisfactory were the: equipment used; size and selection of sample; location and

Exam tip

Do not add lots of new information in your conclusion. Avoid introducing entirely new ideas, theories or data as part of your conclusion. All important data — be they primary or secondary — should have been written about earlier in the report.

Deviation

Emphasise variations (+/-) from a fixed reference point. Typically the reference point is zero but it can also be a target or a long–term average. Can also be used to show sentiment (positive/neutral/negative).

Example uses
Survey opinions, temperature deviations from mean

Spine

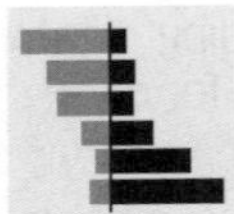

Splits a single value into two contrasting components (eg male/female).

Surplus/deficit filled line

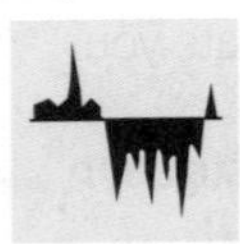

The shaped area of these charts allows a balance to be shown – either against a baseline or between two series.

Correlation

Show the relationship between two or more variables. Be mindful that, unless you tell them otherwise, many readers will assume the relationships you show them to be causal (i.e. one causes the other).

Example uses
Life expectancy and income in different places, rainfall and runoff

Scatterplot

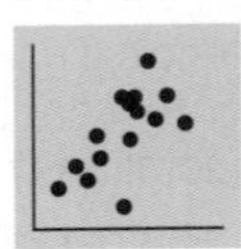

The standard way to show the relationship between two continuous variables, each of which has its own axis.

Bubble

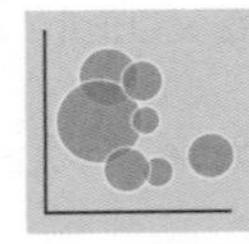

Like a scatterplot, but adds addtional detail by sizing the circles according to a third variable.

Ranking

Use where an item's position in an ordered list is more important than its absolute or relative value. Don't be afraid to highlight the points of interest.

Example uses
Wealth or income in different places, survey opinions, ranking of sediment classes (sand, gravel) at different sites

Lollipop

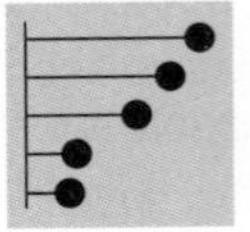

Lollipops draw more attention to the data value than standard bar/column and can also show rank and value effectively.

Slope

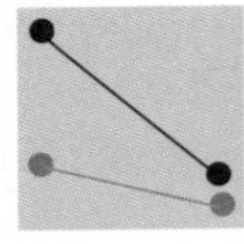

Perfect for showing how ranks have changed over time or vary between categories.

Distribution

Show values in a dataset and how often they occur. The shape (or 'skew') of a distribution can be a memorable way of highlighting the lack of uniformity or equality in the data.

Example uses
Income or population distribution, sediment analysis

Boxplot

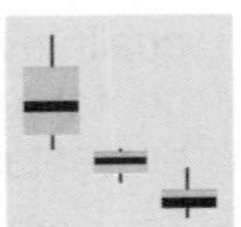

Summarise multiple distributions by showing the median (centre) and range of the data.

Violin plot

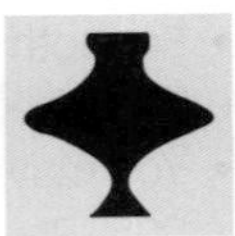

Similar to a box plot but more effective with complex distributions (data that cannot be summarised with simple average).

Magnitude

Show size comparisons. These can be reflective (just being able to see larger/bigger) or absolute (need to see fine differences).

Example uses
Sediment orientation, directional traffic or pedestrian flows

Grouped symbol

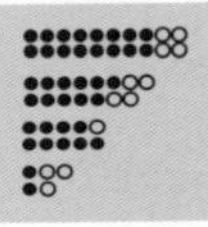

An alternative to bar/column charts when being able to count data or highlight individual elements is useful.

Radar

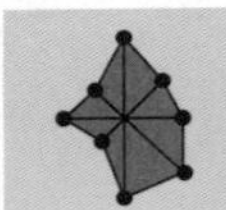

A space-effecient way of showing value of multiple variables — but make sure they are organised in a way that makes sense to reader.

Part-to-whole

Show how a single entity can be broken down into its component elements. If the reader's interest is solely in the size of the components, consider a magnitude-type chart instead.

Example uses
Survey results, wet and dry days in the calender month, prevailing and dominant wind frequency

Donut

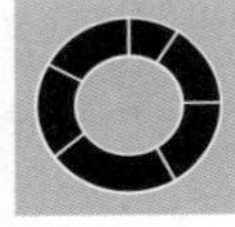

Similar to a pie chart — but the centre can be a good way of making space to include more information about the data (e.g. total).

Gridplot

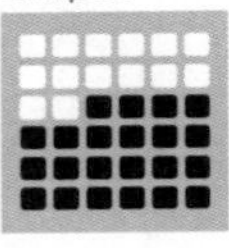

Good for showing % information, they work best when used on whole numbers and work well in multiple layout form.

Change over Time

Give emphasis to changing trends. These can be short (intra-day) movements or extended series traversing decades or centuries.

Example uses
Historical population changes (using census data), temperature and rainfall changes over time

Area chart

Use with care — these are good at showing changes to total, but seeing changes in components can be very difficult.

Calendar heatmap

The great way of showing temporal patterns (daily, weekly, monthly) — at the expense of showing precision in quantity.

Spatial

Aside from locator maps, only used when precise locations or geographical patterns in data are more important to the reader than anything else.

Example uses
Populations or climate data, migration flows, ice flows, house prices

Scaled cartogram (value)

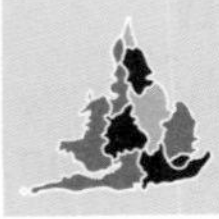

Stretching and shrinking a map so that each area is sized according to a particular value.

Proportional symbol (count/magnitude)

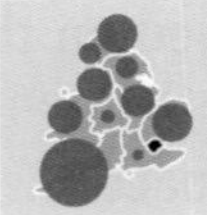

Use for total rather than rates — be wary that small differences in data will be hard to see.

Flow

Show the reader volumes or intensity of movement between two or more states or conditions. These might be logical sequences or geographical loocations.

Example uses
Flows of people, investment, goods or resources between places, physical movement of water or sediment.

Sankey

Show changes in flows from one condition to at least one other; good for tracing the eventual outcome of a complex process.

Chord

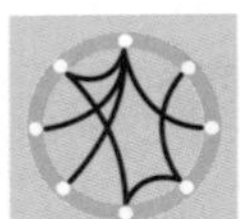

A complex but powerful diagram which can illustrate 2-way flows (and net winner) in a matrix.

Priestley timeline

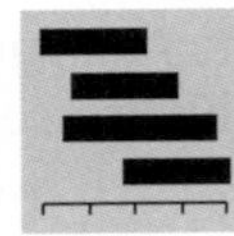

Great when date and duration are key elements of the story in the data.

Circle timeline

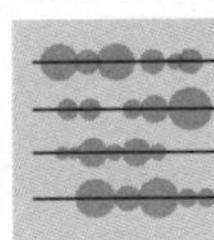

Good for showing discrete values of varying size across multiple categories (e.g. earthquakes by continent).

Flow map

For showing unambiguous movement across a map.

Contour map

For showing areas of equal value on a map. Can use deviation colour schemes for showing +/- values.

Figure 5.2 Examples of different graphical techniques

Reproduced by permission of the *Financial Times*

time of surveys; quality and quantity of data? Try to show that you have thought deeply about potential or real flaws in the methods used. You may even consider how the original fieldwork question or hypothesis could be modified or improved in light of what you have since discovered.

The difference between...

The table below shows two contrasting styles of conclusion and evaluation. In this case, two students have completed a study of variations in urban environmental quality in neighbouring London places. Student B's offering is by far the superior of the two approaches for the following reasons:

- Student A gives descriptive, listed and unconnected statements, whereas Student B synthesises (draws together) different findings to create a more convincing picture.
- Student A does not reflect critically on what is being argued at any stage, whereas Student B is more thoughtful and evaluative.
- Student A does not reflect on the validity and reliability of the findings, whereas Student B does.

Low-quality conclusion and evaluation (Student A)	High-quality conclusion and evaluation (Student B)
In conclusion, my research showed that: • environmental quality was lowest along Streatham High Road • environmental quality was highest in neighbouring Balham My actual figures showed that Streatham received a score of 43 using my environmental quality index (EQI), while Balham reached 103 out of a possible 120 points. This is a very large difference to have discovered. I also found that nowadays house prices vary greatly between the two survey areas. The highest prices were found where EQI scores were also highest. The lowest prices were found where the lowest EQI scores were. I have therefore answered my question — 'To what extent do environmental quality differences cause house prices to vary between neighbouring places in London?' — by showing that they vary a great deal and that this appears to be strongly linked with EQI scores. Overall, I am very pleased with the results of my study. I think that my results were very reliable, although if I had more time it would have been good to carry out a larger sample and to take more readings of environmental quality. I would also design my EQI recording sheet more carefully and make sure all the tick boxes were on a single sheet of paper.	There were significant differences in environmental quality in the two study areas: Balham's EQI scores were notably higher than Streatham's. These differences appeared to be strongly linked with house price variations (and thus demographic differences in terms of average incomes). These conclusions are tentative however because I used semi-qualitative data (the quality index) based on my own personal judgements of quality. My primary data backed up the secondary data findings, which showed air pollution to be highest in Streatham (UCL research) and poverty to be far higher there (2011 Census). Moreover, chi-squared tests on the data showed the results to be Significant (99% confidence limits). However, the patterns were often more complex at the local scale: some postcodes in Streatham have high EQI scores and in Balham some are very low. If time had allowed, I would have liked to explore this scale issue further. During my analysis, however, it became unclear whether the house price variations were really a *cause* or an *effect* of the EQI scores I recorded. It is possible that there are other causal factors explaining differences in EQI scores that I was not able to uncover. If I were to repeat the work I would additionally examine how these places are represented in the media in order to see if that affects who migrates into these places, and the implications of this for house prices and EQI scores.

Final checklist

Table 5.5 provides you with a useful final checklist to follow. Several of these 'good practice' tips deal with the all-important presentation issues.

Table 5.5 A final checklist

Don't forget to...
• Choose a tightly focused fieldwork question and, if relevant, a strictly limited number of hypotheses. • Link the fieldwork question clearly to your geography specification. • Ensure that your proposed work has a clear spatial component, and involves collecting data that, ideally, you can represent on maps. Be careful not to carry out what is essentially a biology or economics investigation. • Make sure that you know about the 'specialised concepts' — such as risk and threshold — and how they can be applied (if unsure, ask your teacher). • Personalise any downloaded maps to show the location, choice of topic and/or sample points, following standard geographical conventions such as including a scale and north arrow. • Justify (in detail) all the methods used and explain the sampling method(s) employed. • Ensure that ample quantitative data have been collected for graphs. Limit the application of statistical tests such as Spearman's rank correlation to situations where sufficient data have been collected. • Be aware of the wide range of graphical techniques and simple statistical tools that are available for your data analysis. Fieldwork investigations benefit when a variety of techniques are used. • Avoid including too many extensive tables in your investigation, especially in the sections for methods and evaluation. • Incorporate a wide variety of appropriate graphical and mapping techniques in your analysis. • In your analysis focus on interpreting (not just describing) results and explaining your findings, highlighting any spatial patterns or trends identified. • Number and place all the illustrations appropriately within the text, and then refer to them throughout the written analysis. • Pay close attention to the assessment criteria and follow the recommended report structure. • Print and present your finished report in a clear and user-friendly format, following your specification's guidelines.

You should know

- **Success in your independent investigation is likely to equate with success in A-level Geography as a whole, so make sure that you prioritise this work.**
- **It is very important to have a clear vision of what you want to achieve very early on in the planning process, and it is essential that you familiarise yourself with the marking criteria.**
- **It is necessary to demonstrate a high level of independence, rather than relying on your teachers to do the thinking for you. That is why it is called an 'independent investigation'.**
- **A successful independent investigation must have a focus that is neither too broad (and therefore unmanageable) nor too narrow to allow for much analysis or evaluation.**
- **In order to gain the highest marks, you must justify each step you take and provide sustained evidence that you are thinking critically about the *entire* research and fieldwork process.**
- **High-quality independent investigations are likely to employ a range of interesting data collection and presentation techniques, possibly making use of new technologies.**

6 Exam board focus

Learning objectives

- To get to know the structure and timings for your own exam board's examination papers
- To become familiar with the types of question asked and their relationship to the assessment objectives (AOs)
- To understand how synopticity is examined
- To become familiar with the unique features of your exam board's assessments, in order to gain the highest grades

AQA A-level

All the information about this specification can be found at:
www.aqa.org.uk/subjects/geography/as-and-a-level/geography-2030

Exam structure and topics

The AQA A-level is made up of two exam papers with a total assessment time of 5 hours (Table 6.1).

The independent investigation is worth 20% of the total marks available.

Table 6.1 AQA A-level Geography structure

Paper 1: Physical geography	Paper 2: Human geography
Duration: 2 hours 30 minutes **Section A: Water and carbon cycles** Data analysis, short-answer tasks and one evaluative essay **Section B: Landscape options** Three options (choose one): • Hot desert systems and landscapes • Coastal systems and landscapes • Glaciated systems and landscapes Data analysis, short-answer tasks and one evaluative essay **Section C: Options** Two options (choose one): • Hazards • Ecosystems under stress Multiple-choice questions, data analysis, short-answer tasks, two 'mini essays' — one of which may involve a connection to another part of the specification (synoptic) — and one evaluative essay	Duration: 2 hours 30 minutes **Section A: Global systems and global governance** Data analysis, short-answer tasks and one evaluative essay **Section B: Changing places** Data analysis, short-answer tasks and one evaluative essay **Section C: Options** Three options (choose one): • Contemporary urban environments • Population and the environment • Resource security Multiple-choice questions, data analysis, short-answer tasks, two 'mini essays' — one of which may involve a connection to another part of the specification (synoptic) — and one evaluative essay

Typical marks and timing

Table 6.2 outlines a typical mark structure and timing for the AQA exams.

Table 6.2 AQA A-level Geography marks and timing

Paper	Questions	Mark	Typical commands (and AO targeting)	Minutes per question*	Refer to
1	1.1, 2.1, 3.1, 4.1	4	Explain (AO1)	5	Chapter 3
	1.2, 2.2, 3.2, 4.3, 5.5, 6.5	6	Analyse, Compare, Assess, Evaluate (AO3)	7	Chapter 2
	1.3, 2.3, 3.3, 4.3	6	Assess (AO1 and AO2)	7	Chapter 3
	5.6, 5.7, 6.6, 6.7	9	Analyse, Assess, Evaluate, To what extent (AO1 and AO2)	12	Chapter 4
	1.4, 2.4, 3.4, 4.4, 5.8, 6.8	20	Assess, Evaluate, To what extent, How far do you agree (AO1 and AO2)	25	Chapter 4
2	1.1, 2.1	4	Explain (AO1)	5	Chapter 3
	1.2, 2.2, 3.5, 4.5, 5.5	6	Analyse, Compare, Assess, Evaluate (AO3)	7	Chapter 2
	1.3, 2.3	6	Assess (AO1 and AO2)	7	Chapter 3
	3.6, 4.6, 5.6, 3.7, 4.7, 5.7	9	Analyse, Assess, Evaluate, To what extent (AO1 and AO2)	12	Chapter 4
	1.4, 2.4, 3.8, 4.8, 5.8	20	Assess, Evaluate, To what extent, How far do you agree (AO1 and AO2)	25	Chapter 4

*These suggested approximate timings do not take into account any extra time allowed for some candidates.

Three unique features of the AQA exams

In order to maximise your AQA grade, make sure that you are familiar with any unique features of the AQA geography exams. In particular, watch out for:

- *The requirement to use your own knowledge when answering some of the resource-based questions.* There is no AO3 'descriptive' credit available for questions 1.3, 2.3, 3.3 and 4.3 (Paper 1) or 1.3 and 2.3 (Paper 2). While there should be some reference to the stimulus material, these particular resources are intended to function more as a 'launch pad' for you to bring in other relevant information that you know about.
- *The 9-mark 'mini essays' on both papers.* It is quite a challenge to apply the lessons learned in Chapter 4 to a 9-mark essay that must be written in just 12 minutes. It is, however, a requirement of the AQA course that you do so! Make sure that the page of writing you produce is broken up into several short, punchy paragraphs and, if possible, includes a brief, evaluative conclusion.
- *The 9-mark synoptic (or 'linking') question.* Pay particular attention to the wording of the 9-mark questions in the exam, one of which could be a synoptic assessment that requires you to draw on both human and physical geography in your response. Alternatively, one of the 20-mark questions could have a synoptic element included.

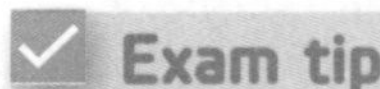

Exam tip

Make sure you are extremely familiar with the specimen examination papers and any past papers that are available. AQA exams use many different command words in varying ways so it is essential that you do not rely solely on the command word to help you remember which AO(s) the question is targeting. 'Assess' can be used for data questions as well as non-data questions, for instance. You will also need to be very familiar with the structure of the examination papers and the ways different AOs are targeted by different parts of the assessment.

Edexcel A-level

All the information about this specification can be found at: http://qualifications.pearson.com/en/qualifications/edexcel-a-levels/geography-2016.html

Exam structure and topics

The Edexcel A-level is made up of three exam papers with a total assessment time of 6 hours 45 minutes (Table 6.3).

The independent investigation is worth 20% of the total marks available.

Table 6.3 Edexcel A-level Geography structure

Paper 1	Paper 2
Duration: 2 hours 15 minutes **Section A: Tectonic processes and hazards** Data analysis, short-answer tasks and one evaluative essay **Section B: Landscape systems, processes and change** Two options (choose one): • Glaciated landscapes and change • Coastal landscapes and change Data analysis, short-answer tasks and one evaluative essay **Section C: Physical systems and sustainability** Data analysis, short-answer tasks and one evaluative essay (focused on the water and carbon cycles)	Duration: 2 hours 15 minutes **Section A: Globalisation/superpowers** Short-answer tasks and two evaluative essays **Section B: Shaping places** Two options (choose one): • Regenerating places • Diverse places Data analysis, short-answer tasks and one evaluative essay **Section C: Global development and connections** Two options (choose one): • Health, human rights and intervention • Migration, identity and sovereignty Data analysis, short-answer tasks and one evaluative essay
Paper 3	
Duration: 2 hours 15 minutes **Synoptic investigation** An extended series of data analysis, short-answer tasks and evaluative essays (based on a previously unseen resource booklet)	

Typical marks and timing

Table 6.4 outlines a typical mark structure and timing for the Edexcel exams.

Table 6.4 Edexcel A-level Geography marks and timing

Paper	Questions	Mark	Typical commands (and AO targeting)	Minutes per question*	Refer to
1	1(a)	4	Draw, calculate (AO3)	5	Chapter 2
	2(a), 3(a)(i), 3(a)(ii), 4(a)(i), 4(a)(ii)	6	Explain (AO1 and AO2)	8	Chapter 3
	2(b), 3(b), 4(b), 4(c)	8	Explain (AO1)	10	Chapter 3
	4(d)	12	Assess (AO1 and AO2)	16	Chapter 4
	2(c), 3(c), 4(e)	20	Evaluate (AO1 and AO2)	26	Chapter 4

→

Paper	Questions	Mark	Typical commands (and AO targeting)	Minutes per question*	Refer to
2	1(a), 2(a), 2(b), 3(b), 4(b) 5(c), 6(c)	4, 8	Explain (AO1)	5, 10	Chapter 3
	3(a)(ii), 4(a)(ii), 5(b), 6(b)	6	Suggest (AO1 and AO2)	8	Chapter 3
	1(b), 2(b)	12	Assess (AO1 and AO2)	16	Chapter 4
	3(c), 4(c), 5(d), 6(d)	20	Evaluate (AO1 and AO2)	20	Chapter 4
3	1(a)	4	Explain (AO1)	8**	Chapter 3
	3, 4	8	Analyse (AO1 and AO3)	16**	Chapter 2
	5, 6	18, 24	Evaluate (AO1, AO2 and AO3)	36, 48**	Chapter 4

*These suggested approximate timings do not take into account any extra time allowed for some candidates.
**Includes reading time.

Three unique features of the Edexcel exams

In order to maximise your Edexcel grade, make sure that you are familiar with any unique features of the Edexcel geography exams. In particular, watch out for:

- *The high number of AO2 marks available for evaluative essays in all three papers* (see Chapter 4).
- *The need to develop good synoptic skills in order to tackle Paper 3.* As part of your Paper 3 answers, you will need to apply a range of knowledge from different topics that you have learned about and also make good analytical use of the previously unseen resource booklet.
- *The difference between 'assess' and 'evaluate'.* There is a high expectation that answers to essays using the command word 'evaluate' will conclude well and reach a judgement. Essays using the command word 'assess' do not necessarily need to reach a strong overall judgement in order to access the highest marks; however, a summative conclusion is probably still a good idea. It is also important to adopt an ongoing reflective style of writing (see Chapter 4).

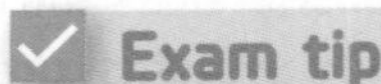

Exam tip

Take care with Paper 3. Make sure you are extremely familiar with the specimen examination papers and any past papers that are available. It is essential that you know which AO(s) each question is targeting. In particular, note that the two long essays target *all three* AOs. To tackle these essays successfully you need to draw on your data analysis skills (Chapter 2) as well as your evaluative essay skills (Chapter 4).

Eduqas A-level

All the information about this specification can be found at:
www.eduqas.co.uk/qualifications/geography/as-a-level

Exam structure and topics

The Eduqas A-level is made up of three exam papers with a total assessment time of 6 hours (Table 6.5). The independent investigation is worth 20% of the total marks available.

Table 6.5 Eduqas A-level Geography structure

Component 1: Changing landscapes and changing places	Component 2: Global systems and global governance	Component 3: Contemporary themes in geography
Duration: 1 hour 45 minutes **Section A: Changing landscapes** Two options (choose one): • Coastal landscapes • Glaciated landscapes Data analysis, short-answer tasks and one evaluative essay **Section B: Changing places** Data analysis, short-answer tasks and one evaluative essay	Duration: 2 hours **Section A: Global systems** Data analysis, short-answer tasks and one evaluative essay (focused on the water and carbon cycles) **Section B: Global governance** Data analysis, short-answer tasks and one evaluative essay (focused on global migration and global governance of the oceans) **Section C: 21st century challenges** One evaluative/synoptic essay using resource material and previously unseen resources	Duration: 2 hours 15 minutes **Section A: Tectonic hazards** One evaluative essay **Section B: Contemporary themes in geography** Four options (choose two): • Ecosystems • Economic growth: India *or* China *or* Development in an African context • Energy challenges • Weather and climate Two evaluative essays (one per option)

Typical marks and timing

Table 6.6 outlines a typical mark structure and timing for the Eduqas exams.

Table 6.6 Eduqas A-level Geography marks and timing

Paper	Questions	Mark	Typical commands (and AO targeting)	Minutes per question*	Refer to
1	1(a)(i), 2(a)(i), 5(a)(i), 6(a)(i)	5	Compare, Contrast, Describe (AO3)	6	Chapter 2
	1(a)(ii), 2(a)(ii)	6	Suggest (AO2)	7	Chapter 3
	5(b), 6(b)	6	Describe, Outline (AO1)	7	Chapter 3
	3, 4, 7, 8	15	Examine, Assess (AO1 and AO2)	20	Chapter 4
2	1(a), 2(a), 5(a), 6(a)	5	Compare, Contrast, Describe (AO3)	5	Chapter 2
	1(b), 2(b), 5(b), 6(b)	5	Describe, Outline (AO1) or Suggest, Explain why (AO2)	5	Chapter 3
	3, 4, 7, 8	20	Evaluate, To what extent, Discuss (AO1 and AO2)	22	Chapter 4
	9, 10	30	Evaluate, To what extent, Discuss (AO1, AO2 and AO3)	3	Chapter 4
3	1–2	38	Evaluate, To what extent, Discuss (all AOs)	40	Chapter 4
	3–12	45	Evaluate, To what extent, Discuss (all AOs)	47	Chapter 4
*These suggested approximate timings do not take into account any extra time allowed for some candidates.					

Three unique features of the Eduqas exams

In order to maximise your Eduqas grade, make sure that you are familiar with any unique features of the Eduqas geography exams. In particular, watch out for:

- *The high number of AO2 marks available for evaluative essays, especially on Paper 2* (see Chapter 4).
- *The requirement for synopticity in optional questions 9 and 10 on Paper 2.* As part of your answer, you need to apply a range of knowledge from different topics.
- *The high number of AO3 marks across Papers 1 and 2.* There are 10 AO3 marks available for each section on each paper (50 in total), so seize every opportunity to study and practise for this (see Chapter 2).

Exam tip

A small number of AO3 skills marks are available for the Paper 3 essays. These are awarded for essays that (i) are presented in a well-structured and logical way and (ii) communicate arguments and conclusions coherently. Take extra care to structure your work carefully into paragraphs, and leave enough time to write a coherent conclusion.

OCR A-level

All the information about this specification can be found at: www.ocr.org.uk/qualifications/as-a-level-gce-geography-h081-h481-from-2016

Exam structure and topics

The OCR A-level is made up of three exam papers with a total assessment time of 5 hours and 30 minutes (Table 6.7).

The independent investigation is worth 20% of the total marks available.

Table 6.7 OCR A-level Geography structure

Paper 1: Physical systems	Paper 2: Human interactions	Paper 3: Geographical debates
Duration: 1 hour 30 minutes **Section A: Landscape systems** Three options (choose one): • Coastal landscapes • Glaciated landscapes • Dryland landscapes Data analysis, short-answer tasks and one evaluative essay **Section B: Earth's life support systems** Data analysis, short-answer tasks and two evaluative essays (focused on the water and carbon cycles)	Duration: 1 hour 30 minutes **Section A: Changing spaces, making places** Data analysis, short-answer tasks and one evaluative essay **Section B: Global connections** Two options (choose one): • Trade in the contemporary world • Global migration Data analysis, short-answer tasks Two further options (choose one): • Human rights • Power and borders One evaluative essay	Duration: 2 hours 30 minutes **Sections A, B and C** Five options (choose two): • Climate change • Disease dilemmas • Exploring oceans • Future of food • Hazardous Earth Section A: short-answer tasks Section B: two evaluative essays (one per chosen option) Section C: two evaluative essays (one per chosen option)

Typical marks and timing

Table 6.8 outlines a typical mark structure and timing for the OCR exams.

Table 6.8 OCR A-level Geography marks and timing

Paper	Questions	Mark	Typical commands (and AO targeting)	Minutes per question*	Refer to
1	1(a), 2(a), 3(a)	8	Explain (AO1)	11	Chapter 3
	1(b), 2(b), 3(b)	4	Calculate (AO3)	6	Chapter 2
	1(c), 2(c), 3(c)	3	Explain (AO2)	5	Chapter 3
	4(c)	10	Examine (AO1 and AO2)	14	Chapter 4
	1(d), 2(d), 3(d), 4(d)	16	Evaluate, To what extent, Discuss (all AOs)	22	Chapter 4
2	1(b)	8	Explain why (AO2 and AO3)	11	Chapter 3
	1(c), 2(b), 3(b)	6, 8	Explain (AO1)	9, 11	Chapter 3
	2(a)(i), 3(a)(i), 2(a)(iii), 3(a)(iii)	2, 4	Suggest (AO3)	3, 6	Chapter 2
	1(d)	16	Evaluate, To what extent, Discuss (all AOs)	22	Chapter 4

3	1–5 (part a) 1–5 (part b) 6–10 11–15	3 6 12 33	Identify (AO3) Explain (AO1) Assess, Examine, To what extent (AO1 and AO2) Assess, Discuss, To what extent (AO1 and AO2)	5 9 16 46	Chapter 2 Chapter 3 Chapter 4 Chapter 4
*These suggested approximate timings do not take into account any extra time allowed for some candidates.					

Three unique features of the OCR exams

In order to maximise your OCR grade, make sure that you are familiar with any unique features of the OCR geography exams. In particular, watch out for:

- *The high number of evaluative marks available in essays, especially in Section C of Paper 3.*
- *The requirement for explicit synopticity in Section B of Paper 3.*
- *The requirement to engage with a wide variety of resources in all three papers, such as graphs of various types, photographs, maps (including OS), text passages and cartoons.*

Exam tip

While you are not required to produce long descriptions of case studies, make sure that your responses, especially for the longer essays, are rooted in real-world geography. Paper 3 (Geographical debates) is focused on contemporary issues relevant to today. This suggests that it is a good idea not to rely heavily on textbooks, and to additionally use case studies drawn from today's headlines.

WJEC AS/A-level

All the information about this specification can be found at: www.wjec.co.uk/qualifications/geography/r-geography-gce-asa-from-2016

Exam structure and topics

The WJEC A-level is made up of:

- ➜ Unit 1 and Unit 2 (AS)
- ➜ Unit 3 and Unit 4 (A-level)

The total assessment time is 7 hours and 30 minutes (Table 6.9).

The independent investigation is worth 20% of total marks available.

Table 6.9 WJEC A-level Geography structure

Unit 1: Changing landscapes	Unit 2: Changing places
Duration: 2 hours **Section A: Changing landscapes** Two options (choose one): • Coastal landscapes • Glaciated landscapes Data analysis and short-answer tasks **Section B: Tectonic hazards** Data analysis and short-answer tasks	Duration: 1 hour 30 minutes **Section A: Changing places** Data analysis and short-answer tasks **Section B: Fieldwork investigation** Data analysis and short-answer tasks
Unit 3: Global systems and global governance	**Unit 4: Contemporary themes in geography**
Duration: 2 hours **Section A: Global systems** Data analysis, short-answer tasks and one evaluative essay (focused on the water and carbon cycles) **Section B: Global governance** Data analysis, short-answer tasks and one evaluative essay (focused on global migration and global governance of the oceans) **Section C: 21st century challenges** One evaluative/synoptic essay using resource material and previously unseen resources	Duration: 2 hours **Section A: Tectonic hazards** One evaluative essay **Section B: Contemporary themes in geography** Four options (choose two): • Ecosystems • Economic growth: India *or* China *or* Development in an African context • Energy challenges • Weather and climate Two evaluative essays (one per option)

Typical AS marks and timing

Table 6.10 outlines a typical mark structure and timing for the WJEC AS exams.

Table 6.10 WJEC AS geography marks and timing

Paper	Selected questions	Mark	Typical commands (and AO targeting)	Minutes per question*	Refer to
1 (**bold**) 2 (*italic*)	1(a)(i), 2(a)(i), 3(a)(i), **4(a)(i), 5 (a)(i)**	3, 5	Compare, Identify, Describe (AO3)	3, 5	Chapter 2
	1(a)(ii), 2(a)(ii), 3(a)(ii), **4(a)(ii), 5(a)(ii)**	3, 9	Suggest, Interpret (AO2)	3, 9	Chapter 3
	1(b), 2(b), **3(b), 4(b)**	8	Examine, Assess (AO1 and AO2)	8–9	Chapter 4
	4, 5	9	Evaluate, To what extent, Justify (all AOs)	10	Various
	7(b)	10	Describe, Outline (AO1)	10	Chapter 3

*These suggested approximate timings do not take into account any extra time allowed for some candidates.

Typical A-level marks and timing

Table 6.11 outlines a typical mark structure and timing for the WJEC A2 exams.

Table 6.11 Edexcel A-level Geography marks and timing

Paper	Questions	Mark	Typical commands (and AO targeting)	Minutes per question*	Refer to
3	1(a), 2(a), 5(a), 6(a)	3, 5	Compare, Contrast, Describe (AO3)	4, 6	Chapter 2
	1(b), 2(b), 5(b), 6(b)	4, 5	Describe, Outline (AO1) *or* Suggest, Explain why (AO2)	5, 6	Chapter 3
	3, 4, 7, 8	18	Evaluate, To what extent, Discuss (AO1 and AO2)	23	Chapter 4
	9, 10	26	Evaluate, To what extent, Discuss (all AOs)	34	Chapter 4
4	1–2	20	Evaluate, To what extent, Discuss (all AOs)	38	Chapter 4
	3–12	22	Evaluate, To what extent, Discuss (all AOs)	41	Chapter 4

*These suggested approximate timings do not take into account any extra time allowed for some candidates.

Three unique features of the WJEC exams

In order to maximise your WJEC grade, make sure that you are familiar with any unique features of the WJEC geography exams. In particular, watch out for:

- *The high number of AO2 marks available for evaluative essays, especially in the A2 papers* (see Chapter 4).
- *The requirement for synopticity in optional questions 9 and 10 in Unit 3*. As part of your answer, you need to apply a range of knowledge from different topics.
- *The high number of AO3 marks across Units 1, 2 and 3*. Seize every opportunity to study and practise for this (see Chapter 2).

Exam tip

A small number of AO3 marks are available for the Unit 4 essays. These are awarded for essays that (i) are presented in a well-structured and logical way and (ii) communicate arguments and conclusions coherently. Take extra care to structure your work carefully into paragraphs, and leave enough time to write a coherent conclusion.